# Discovering Evolut█████████gy

# Discovering Evolutionary Ecology

## Bringing together ecology and evolution

Peter J. Mayhew
*University of York, UK*

OXFORD
UNIVERSITY PRESS

# OXFORD

UNIVERSITY PRESS

Great Clarendon Street, Oxford OX2 6DP

Oxford University Press is a department of the University of Oxford.
It furthers the University's objective of excellence in research, scholarship,
and education by publishing worldwide in

Oxford  New York

Auckland  Cape Town  Dar es Salaam  Hong Kong  Karachi
Kuala Lumpur  Madrid  Melbourne  Mexico City  Nairobi
New Delhi  Shanghai  Taipei  Toronto

With offices in

Argentina  Austria  Brazil  Chile  Czech Republic  France  Greece
Guatemala  Hungary  Italy  Japan  Poland  Portugal  Singapore
South Korea  Switzerland  Thailand  Turkey  Ukraine  Vietnam

Oxford is a registered trade mark of Oxford University Press
in the UK and in certain other countries

Published in the United States
by Oxford University Press Inc., New York

British Library Cataloguing in Publication Data
Data available

Library of Congress Cataloging in Publication Data
Data available

Typeset by Newgen Imaging Systems (P) Ltd., Chennai, India
Printed in Great Britain
on acid-free paper by
Biddles Ltd., King's Lynn

ISBN  0–19–857060–0         978–0–19–857060–8
ISBN  0–19–852528–1 (Pbk.)  978–0–19–852528–8 (Pbk.)

10 9 8 7 6 5 4 3 2

To Grandpa, A.H. 'Peter' Dunn (1908–2003),
who bestowed on others his love of nature.

# Preface

There's more to this life than just living.

Frank Borman, Apollo 8 astronaut

The natural world is a place I escape to: a place that goes about its business regardless of everyday individual human concerns. It is a place of beauty, change, diversity, and endless fascination. Like many who share these sentiments, I was never content to just be in nature: I had to watch, name, learn, and understand. This book is about understanding how and why the natural world works, thereby to appreciate it more for what it really is. For me, that is one of the things that make life 'more than just living'.

For naturalists, two fields of science feel especially comfortable: ecology and evolution. Ecology is traditionally a science of the great outdoors, dealing with the interactions between organisms and their environment (including other organisms). Evolution is traditionally a science of museum specimens, dealing with how lineages of organisms arise, change, and eventually go extinct. Both ecologists and evolutionary biologists share a common goal: they want to understand the diversity of life; how it arises, how it is maintained, and why sometimes it is not. They should have a lot to say to each other. The field where ecologists and evolutionary biologists meet is called evolutionary ecology and, despite having 150-year-old roots, it has only recently matured into something that can fill books.

This book has one overriding aim: to synthesize the field of evolutionary ecology; that is, to explain what the field as a whole has discovered, rather than just all the little bits. Along the way there is some detail; the work of scientists. While the detail can exist without the synthesis, the synthesis gives the detail added value. While some of the detail may change, be lost, or added to, the synthesis I hope will remain.

I have written primarily for the students of biology whom I meet at undergraduate level. In 1998, as a new lecturer at the University of York, my colleague Richard Law invited me to take over his lectures on evolutionary ecology. However, I found no books that dealt with the field in the way I needed and decided to write my own. I have written the book that I would have wanted as a student: using a short, informal style, so some people might get to the end. As a result this is not a compendium of evolutionary ecology knowledge. There is always more detail in the world, or indeed in any scientific field, than any one person can assimilate. From what little detail we do have, however, we mortals must formulate pictures of the world that we can apply to novel situations, of which the world is full. I hope this book has just enough to do that. The book may also be more widely accessible than I originally meant it to be. I hope that postgraduates and other researchers in the field, who tend to stay within the bounds of a single chapter, will find it useful to have an overall view that places their work in a broader context. The public at large should also have a fighting chance, and I have tried to make that more likely by including a glossary of the more technical terms. Terms included in the glossary appear in bold on first mention.

The precise content of the book was shaped by three secondary desires. First, I did not want to write yet another behavioural ecology book. But, because most evolutionary ecologists study behaviour, if I had devoted space in proportion to the amount of work carried out in the various subdisciplines of the field, that is pretty much what would have happened. However, a behavioural ecology book would not have achieved my broader aims. Instead, I have tried to cover a wide range of topics to do justice to the breadth of the field in ways that previous books have not. Each chapter serves merely as an introduction to each topic, about which others have written entire books. For those who feel like learning a bit more, I make a few recommendations for further reading at the end of each chapter. Some of the topics in the book are not normally considered to lie in evolutionary ecology, but more solidly in mainstream evolution or ecology. I have included them because I feel they should be here.

Second, I am aware that most biologists express a greater enthusiasm for some organisms than others. They spend a lot of time trying to persuade each other that their study organisms are the most interesting. I believe that to appreciate evolutionary ecology to the full, you must be prepared to discard taxonomic and functional prejudice. This does not mean that you should not feel a special affection for some taxa; rather you should not feel disaffection for other taxa. The reader should be prepared for a good mix of the botanical, microbial and zoological, aquatic and terrestrial. To emphasize this even more I have occasionally employed positive discrimination in my choice of material.

Third, I have not made a special effort to emphasize applied questions. Evolutionary ecology can help solve many problems that beset our planet and our species, but my desire here is to help people to love the subject, and not to plague them with worry or guilt. I have included applied questions simply where they provide a fascinating perspective that improves understanding. As it turns out, there should be enough applied biology to keep enthusiasts happy.

The chapters should preferably be read in sequence from start to finish since they build upon each other to provide the overall picture at the end. Because I still wanted this book to be scientific, factual statements are supported by citations from the primary scientific literature, though space and flow limited the extent to which I could do this. Space limitations also meant that I often had to reduce long complicated stories to a few salient points, leaving out alternative viewpoints. This makes it virtually certain that researchers in the field, and possibly other readers, will disagree with me at least once somewhere in the book. I hope that you all find such moments stimulating.

Many people helped in the creation of this book. Biology students at York made comments on my teaching that shaped the way the book was written. Several people, mostly anonymously, reviewed the initial proposal, and I am grateful to all of them. I particularly thank Brian Husband, who convinced me that speciation mechanisms had to be included. I am grateful to the following persons for commenting on draft chapters: Peter Bennett, Calvin Dytham, Ian Hardy, Richard Law, Geoff Oxford, Jeremy Searle, Ole Seehausen, and Mark Williamson.

Permission to reproduce photographs was generously provided by John Altringham, Craig Benkman, May Berenbaum, Didier Bouchon, Sarah Bush, David Conover, James Cook, Angela Douglas, Andrew Forbes, Richard Fortey, Niclas Fritzén, Leslie Gottlieb, Peter Grant, Angela Hodge, Greg Hurst, Mike Hutchings, Ian Hutton, Eric Imbert, Colleen Kelly, E. King, Hans Peter Koelewijn, Thomas Ledig, Mark Macnair, James Marden, Stephane Moniotte, Camille Parmesan, Olle Pelmyr, Thomas Ranius, Loren Rieseberg, Dolph Schluter, Ole Seehausen, Kim Steiner, Robert Vrijenhoek, Truman Young, Arthur Zangerl, and Gerd-Peter Zauke.

I am grateful to the following for permission to reproduce various figures: The American Association for the Advancement of Science, The Royal Society of London, The Society for the Study of Evolution, and Springer Science and Business Media. Ian Sherman at Oxford University Press opened the door to what you are reading, gave valuable advice, displayed admirable patience, and was above all a friendly face. I am grateful to Alastair Fitter, for granting me the sabbatical term in which I made the majority of progress. I was also supported by my colleagues at York who bore the brunt of my 'normal' work while I was on sabbatical, particularly Calvin Dytham and Dale Taneyhill. Finally, thanks to my wife Emese and daughters Alice and Lara, the former for understanding my need to write the book and supporting me in the struggle, and the latter for illustrating to me at first hand many of the interesting concepts mentioned in the book.

# Contents

# 1 Where two fields meet

A teacher of mine once simplified his complex family history by saying that he, like all of us, originated from Olduvai Gorge in Tanzania (the 'cradle of mankind'). Tropical Africa has been a cauldron of diversity not only for our own species. It is, to take one example, surprisingly fishy. The Great Lakes of East Africa (Figure 1.1), and surrounding rivers, contain a whopping 1500 species in just one fish family, the cichlids, familiar to freshwater aquarium enthusiasts. This makes cichlids the most species-rich family of vertebrates, beating such diverse and familiar groups as songbirds and mice. They are so diverse that many still await proper scientific description, and many more are doubtless completely undiscovered. Lakes Victoria and Malawi each contain about 500 species, and about 250 species are found in Lake Tanganyika. Diversity of this sort is what makes our planet such an interesting place, and of course, we have to find out what caused it.

The cichlid species of the East African lakes have not each immigrated there from the surrounding habitat; they were born there, and in most cases they are endemics, being found in just one of the lakes (Fryer and Iles 1972). They are a 'radiation' of species. This radiation is all the more remarkable when the ages of the lakes are considered. Lake Tanganyika is the oldest (but has fewest species) at about 10 million years. Lake Malawi, the second oldest is a mere 1–2 million years old. Lake Victoria, amazingly, may have been completely dry around 14,500 years ago, the end of the last ice age. Since then, 500 cichlid species have been born. If species arose in a clockwork linear fashion, that would mean one new species of fish every 29 years!

The varied lifestyles of the fish are equally impressive. In Lake Victoria, for example, have been found cichlids with the following diets: adult fish, fish larvae, fish scales, fish parasites, freshwater snails, insect and other invertebrate larvae, plant and animal plankton, algae growing on rocks, and vascular plants, all with specialized jaws to match (Figure 1.2). The most impressive radiations have occurred among the 'haplochromine' cichlids living on rocky shores in Lakes Victoria and Malawi (Kocher 2004). Clearly, we need to know how so many species could have formed in such a short time span, why it happened here, why cichlids, and why haplochromines most of all? At stake is our understanding of species richness itself.

**Fig. 1.1**    Seen from space, the Great Lakes of the East African Rift Valley are major landscape features. The two largest ones here are Lake Victoria (right) and Lake Tanganyika (bottom)—Lake Malawi is off the bottom of the picture. Lake Victoria is about 300 km across and its northern tip is on the equator. Photo from the NASA Visible Earth image archive. Black lines indicate national boundaries.

# 1.1   Alternative mechanisms

First, let us think briefly about how species are supposed to form. The standard dogma is that this happens through geographic separation and subsequent differentiation. One lineage splits into two distinct ones because a spatial separation occurs, either through a dispersal event to an isolated

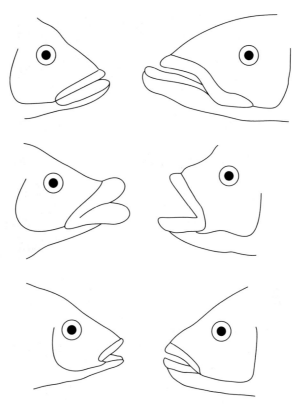

**Fig. 1.2**    The diversity of jaw morphology of Lake Victoria cichlids. Clockwise from top left they eat, snails, fish, fish larvae, algae on rocks, invertebrates on rocks, insect larvae.

new region, or through fragmentation of an existing one (vicariance). The lineages evolve in isolation, through natural selection or other processes, and eventually become distinct enough to be called new species. The differences between related, but geographically isolated species are what gave Darwin and Wallace many clues to their theory of evolution.

Could such processes be at work in the fastest vertebrate radiation? Geographic separation and natural selection have undoubtedly contributed, and a number of observations on geographic distribution and morphological divergence among species are consistent with the process. For example, closely related sister species in Lake Victoria sometimes have widely separated geographic ranges (Seehausen and van Alphen 1999); and different populations of the same species have distinct jaw morphologies that match local diets, suggesting local adaptation (Bouton *et al.* 1999). But there remains a dearth of special explanation: why here and why haplochromines? A growing weight of evidence suggests a role for additional mechanisms and in particular in haplochromines.

What additional mechanisms might be important? Can speciation, for example, occur without geographic isolation? There are two problems that need to be overcome. First, there has to be ecological divergence: the two incipient species have to occupy different niches to prevent them from competing and allow stable coexistence. Second, there has to be reproductive divergence, so that interbreeding does not occur. Getting these events to occur without geographic isolation is a conceptual challenge that has long occupied evolutionary biologists. In the 1990s, this question was bothering cichlid enthusiast, Ole Seehausen. Ole's hunch was that species could diverge *in situ* into reproductively isolated populations by assortative mating based on male coloration. Over time, mate selection by different females for different coloured males would produce two reproductively isolated species living in the same ecological niche but differing in male coloration. Once separated like this, the way would be open for natural selection to allow niche differentiation. The process could then repeat itself. The power of this mechanism is its potential speed. Initial ecological differentiation need only be small, and the constant disruptive power of female choice would drive populations rapidly apart. It was a process that seemed capable of giving rise to a multitude of species in a very short time.

What evidence supported this hypothesis? One source is patterns of geographic overlap between species. If speciation has occurred in the absence of geographic separation, there should also be groups of closely related species that overlap in range a lot. In fact, there are many such cases in Lake Victoria (Seehausen and van Alphen 1999). What about sexual selection? In the field, **sympatric** sister species tended to be opposite colours more commonly than allopatric pairs of species. This is consistent with the origination of new species via selection on coloration *in situ*. These patterns have also recently been demonstrated in Lake Malawi cichlids (Allender *et al.* 2003). In the laboratory, females from red species behaved preferentially towards red males, as did females of blue species towards blue males. When exposed to monochromatic light that hid the males' bright nuptial hues, females would no longer show a mate preference (Seehausen and van Alphen 1998). This was indeed assortative mating based on colour. But why should female mate choice be disruptive? One possible answer is perceptual bias: the colour-sensitive cone cells of haplochromines are particularly sensitive to red and blue parts of the spectrum, and these different sensitivities could lead females to perceive red or blue males preferentially (Seehausen *et al.* 1997). However, other possible mechanisms could be at work. What ever the mechanism, female haplochromines agree with Winston Churchill when he said: 'I cannot pretend to feel impartial about colours. I rejoice with the brilliant ones and am genuinely sorry for the poor browns'.

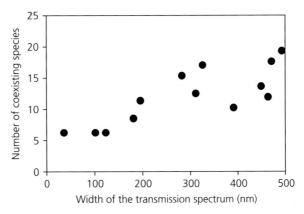

**Fig. 1.3**    Number of coexisting cichlid species against the clarity of the water at different sites in Lake Victoria (a wide transmission spectrum represents clear water). After Seehausen *et al.* (1997), with permission from AAAS.

Another feature of haplochromine cichlids is that if mating does take place between individuals of different species, the offspring are normally perfectly viable. The only reason they can be called separate species at all is because of their fussy mate preferences. Evolutionary biologists call this 'pre-zygotic' isolation. In the field, Ole started to find rather disconcerting observations that mimicked what he was seeing in the laboratory (Seehausen *et al.* 1997). Where the water was murky, and that was often quite a recent phenomenon, he found few species of fish (Figure 1.3) and of dull brown coloration. In clear waters, many species coexisted together, and they were beautifully coloured. It looked as if previous mating barriers were breaking down. Turn it on its head, and mate choice in clear water seemed to have allowed divergence and maintenance of species in the first place.

Could disruptive mate choice be the reason why it is the cichlids, and not some other fish group, that have diverged in this way, and especially the haplochriomine fish that radiated in lakes Victoria and Malawi? That too appears to be the case. Comparing the incidence of mating system and male nuptial coloration in different cichlid groups, Ole showed that there was a significant association between the incidence of polygyny (where males mate with more than one female, long associated with highly selective female mate choice) and male nuptial coloration. Furthermore, the base of the radiation that gave rise to the fish 'superflocks' of Lake Victoria and Malawi, the haplochromines, was characterized by the origin of male nuptial coloration (Seehausen *et al.* 1999).

Could not some other fish group possessing strong sexual selection also have radiated? Put another way; is there anything else about the cichlids, which would lead to this mating system being particularly diversifying for them? Part of the answer may have to do with that second essential process of speciation without geographic isolation, ecological divergence. Some kind of novel ecological flexibility might open up new niches, making each new speciation experiment more likely to succeed. In fact cichlids have long been known to possess a novel character that would lead to such flexibility: the 'decoupled pharyngeal jaw' apparatus (Liem 1973). The bones of the mouth have been freed to evolve into specialized food-gathering implements, while the bones at the back have become very efficient grinding elements. This novelty has given the cichlids jaw-evolvability as well as behavioural plasticity. That it has played an important role in the present diversity of cichlids is a very good bet.

Therefore, much evidence points towards a role for disruptive sexual selection acting on male coloration, followed by ecological differentiation as the reason why cichlids, and particularly those in Lakes Victoria and Malawi, have diversified so rapidly and why those species are still maintained. It is a nice idea. But does a world with those simple conditions produce the desired result? Will it also work in theory? This problem was tackled by a talented undergraduate at the University of Utrecht, Sander van Doorn, who built a simulation model of the process (van Doorn *et al.* 1998). This step is an important one, because ultimately biologists want to put aside a small set of essential processes into a body of theory that captures the essence of reality. We have to know what processes are sufficient and important, and which are just noise.

## 1.2 Simulated lakes and simulated radiations

Theoretical models consist of assumptions, a best guess about how things work in nature, and predictions, which are the model results. A good model will make a few biologically reasonable assumptions and result in predictions that bear a strong resemblance to reality, hence isolating the important mechanisms.

Van Doorn and colleagues started by assuming that individual fish can be characterized by a colour preference (of females), a pigmentation (of males), and by their niche use (represented for simplicity by a single number: think of it as prey size, or water depth). Individuals compete, and are more likely to die if their niche use is similar to that of other individuals. This keeps the population size limited. Fish are born by sexual reproduction, which is dependent on female mate preference, male pigmentation, and the degree of niche overlap (similar niches increases the probability of mating). Mate preference, male pigmentation, and niche use are also heritable, so that offspring resemble their parents, but imperfectly, so small random changes (mutations) are created in each generation. Finally, the more brightly coloured males are, the lower their survival as a result of natural selection (such as predation). So far, so good.

One other important assumption is that females have peaks in perceptual ability at both ends of the colour spectrum. Perceptual ability relates the pigmentation of males to the colour perceived by females. In perfectly clear water, there is a near-perfect match between the two, although females perceive very bright pigments (at either end of the colour spectrum) slightly better than others. In very murky water, all pigments appear brown to females. In slightly murky water, only pigments that are close to female's perceptual peaks are perceived to be coloured.

The model was run by starting off a small population of a single species in clear water and letting mating, reproduction, and death take its course. Species were defined as groups of individuals that, because of their niche, colour, or preference, were very unlikely to mate, and could hence evolve independently of the others. After 2000 generations, five species were coexisting from the original species in this simple and tiny virtual lake. The process could clearly work. How exactly does it happen?

The key is female mate choice. As a result of the biased perception of red and blue, on average females prefer males that have more-extreme-than-average pigmentation. As a result, both pigments and preference become more extreme over time (Figure 1.4). The process is a familiar concept in

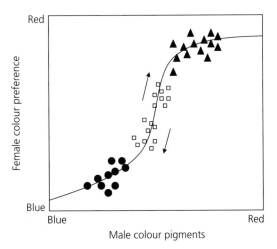

**Fig. 1.4** Speciation via sexual selection in the van Doorn *et al.* (1998) model. Individuals of different species are represented by different symbols. The curved line represents female preference for male colour and is biased towards red and blue (females on average prefer males that are bluer or redder than the population average). Because of this bias, brown species gradually split into two, one redder and one bluer, as can be seen with the species represented by the open squares. After van Doorn *et al.* (1998), with permission from the Royal Society of London.

sexual selection theory and is known as a 'runaway' process. Species that have neutral colours and preferences, neither red nor blue, will split into two species with slightly brighter (redder or bluer) colours. Once incipient species no longer interbreed, their niches diverge as a result of competition (this can not happen in a single species because interbreeding stops the niche changing). The amount of niche space present limits the number of species that can coexist, and it is for this reason that the model only produces a few species. If species have different niches, they can have more similar colours without losing their integrity as species. That of course is exactly what we see in Lake Victoria: lots of species with the same nuptial colour.

The final triumph of the model is what happens when the water is made turbid. Species cannot diverge, or any longer remain sexually isolated because all males appear the same to females. Species number crashes, just as in nature. The model is successful because by using the small pieces of biology gathered so far, it successfully predicts many of the important patterns in nature: it is a good conceptual cartoon for what goes on in nature.

However, the model appears not to be the last word in cichlid speciation. Species in the model form from brown fish gradually splitting into slightly less brown ones. In fact, individual species in nature often display a male red/blue colour **polymorphism**, suggesting that speciation and colour change are much more instantaneous. Thus, the model is in some respects only a rough cartoon of some of the actual processes. In addition, there is a second type of colour polymorphism within some species in which females vary in colour and are associated with a rather interesting genetical system (Seehausen and van Alphen 1999; Seehausen *et al.* 1999). Something different must be going on in those.

Teaming up with theoretician Russ Lande, Seehausen devised a model that incorporates these 'instant' novel female colour morphs with the strange genetics in a sympatric speciation scenario (Lande *et al.* 2001). They showed that given the way novel colour morphs and other traits are inherited together, rapid speciation is likely to result even without ecological differentiation. The female colour polymorphism is due to a gene that causes sex reversal from male to female and is associated with a distinct colour pattern (Seehausen *et al.* 1999) (Figure 1.5).

Imagine then that novel colours are only seen in females. Unusual males that prefer, or do not discriminate against, this colour now have high mating success for two reasons; they are rare male **phenotypes**, so get all the mating with unusual coloured-females that normal males pass by. In addition, if the sex-reversal gene is widespread, they will also be the rarer sex, so get more mates anyway. This process, which favours the novel males through rarity of the male sex, is called sex ratio selection. We will encounter this process again in Chapter 5. An association between the new colour morph and preference

**Fig. 1.5**   A cichlid, *Paralabidochromis chilotes*, from Lake Victoria (length 15 cm). Blotchy morphs like these are, in most populations, female, and include sex reversed males that may play a role in speciation by sex ratio selection. Photo courtesy of Ole Seehausen.

for that colour morph builds up. Over only a few dozen generations, a new reproductively isolated species has arisen *in situ*.

It appears likely that at least two *in situ* processes can account for colour-diverse haplochromine species richness: sexual selection and sex ratio selection. Both these processes can cause speciation with geographic separation, but they can also do it in the absence of geographical separation. The processes appear bizarre and extraordinary at first sight. However, both processes are not unexpected in a wider context; we will come across them again later in the book. What then has the cichlid story taught us?

## 1.3   Cichlids and evolutionary ecology

The cichlid story illustrates many of the broader features of evolutionary ecology, the science that involves both ecological and evolutionary knowledge. Evolutionary biology is the field concerned with understanding how biological lineages change through time (anagenesis), split (cladogenesis), and ultimately go extinct. Ecology is concerned with the interaction of organisms with their environment. The organisms can be considered at various levels of a hierarchy, comprising the individual, the population (groups of individuals of the same species), and the community (groups of interacting populations from different species). Communities in turn comprise the **biotic** component of ecosystems, which also include their interactions with the **abiotic** world. Ecology asks how individuals behave in different environments, what determines population size, and the properties of communities and ecosystems, such as their diversity. Knowing all this, why do ecology and evolution interact and how do they do so?

A basic answer, and one that does not require much in-depth study, is that both fields are concerned with understanding similar characteristics. For example, both evolutionary biologists and ecologists would consider species richness as one of the key variables they want to understand. Both too would want to understand why species richness varies across environments, such as different lakes in the case of cichlids, and across **clades**, such as haplochromines versus other cichlids or cichlids versus other fish.

Another answer, that requires some knowledge of the subject, is that evolutionary and ecological processes affect each other (Figure 1.6). They do this in many ways: one way is through adaptation. Darwin's and Wallace's greatest discovery was an understanding of the way in which this occurs: evolution through natural selection. Organisms vary in form (phenotype). These forms are heritable because of variation in their under- lying genetics (genotype). The phenotypes interact with their environment, and some are more successful than others for a variety of reasons: they may survive or reproduce better. This differential success is called natural selection. Thus, the individuals that contribute to the gene pool of the next generation are a subset of those that were born and will pass on that subset of characteristics to the next generation through their genotype. In this way the population changes through time. A second type of selection process is normally distinguished from natural selection: sexual selection. Sexual selection causes evolution of traits affecting mating success in males and females. Both natural selection and sexual selection come about from phenotypes interacting with their environment, and for this reason selection is generally viewed as an ecological process. Natural selec- tion is responsible for the evolution of traits, such as cichlid jaw shape, which governs their ecological niche. Sexual selection is responsible for traits, such as the bright male coloration of cichlids, that influence their mating success.

So ecology, through the medium of selection, causes anagenesis, evolution within lineages. Ecology can also influence cladogenesis, the other big evolu- tionary process. We saw this in the role that water clarity plays in speeding up or slowing down rates of cichlid speciation and extinction.

Evolution can also affect ecology, and this interaction occurs at several levels of the ecological hierarchy (Figure 1.6). If you know about the envir- onment, you can sometimes accurately predict, or at least in retrospect understand, the phenotypes that are favoured. Evolutionary biologists need to do this routinely, and it will be a repeated theme throughout this book. Ecology at the level of the individual is largely concerned with trying to predict how individual traits should be related to the environment through selection pressure. Behavioural ecology is the field that asks what behaviours would suit particular environments, such as the mate preferences seen in haplochromine cichlids. It is one of the richest parts, but only one of

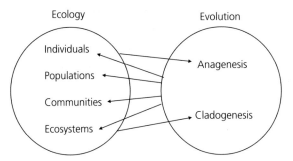

**Fig. 1.6**    The interaction of ecology and evolution.

the parts of evolutionary ecology. Hence evolution by natural selection affects ecology at the level of the individual.

The traits that evolve within species are often relevant to population and community processes. For example, each species has a characteristic reproductive rate, size, and length of life. These are important character-istics in determining how many individuals of a species can exist in any one place, and how variable the populations are. Species also vary in their ecological specialization; for example, how many other species they eat or which eat them. Haplochromine cichlids, for example, have very specialized jaws. These interactions evolve through natural selection, but they also structure communities. Knowing about one should help us to understand the other.

The second major evolutionary process, cladogenesis is also important for an understanding of ecology. To produce species-rich communities, such as in East African lakes, species have to be formed and not go extinct. Both evolution within lineages and the origin and death of lineages are processes that might have contributed. Thus evolution influences every level of the field of ecology and maybe key to understanding some of the basic ecological properties of our planet.

In the following chapters, we will explore the ways in which the two fields of ecology and evolution interact, see what we have learnt about the world as a result, and along the way build up a picture of how exactly these interac-tions occur. However, the book will describe something else about evolu-tionary ecology that cannot be fully appreciated without an overall view of the field. It is that the topics which the field addresses are mutually sup-portive, such that understanding of one aids understanding of others. For example, we can understand the rates of speciation in cichlids from a knowl-edge of speciation and extinction mechanisms, and we can understand those from a knowledge of sexual selection and sex determination. Ultimately then, workers in one area will benefit from an awareness of other areas. This is what makes a synthesis worthwhile. Knowing how these interactions between topics occur reveals interesting features about how our living universe is shaped, and provides another aspect to the bigger picture that the field depicts. The next chapter looks at how organisms became complex from very simple beginnings.

## 1.4   Further reading

The arguments here about cichlid speciation are well described in Seehausen (2000), and much of the story is told in Seehausen *et al.* (1997) and the Galis and Metz (1998) commentry on this. Meyer (1993) and Turner (1999) are also useful. More general reviews about cichlids, including speciation and sexual selection, are in Kornfield and Smith (2000) and Kocher (2004). The general issue of **sympatric speciation** is reviewed by Via (2001).

# 2  Evolutionary cover-stories

> Great things are not done by impulse, but by a series of small things brought together.
>
> Vincent van Gogh

People have long suspected that the first organisms must have been relatively simple. Since the origin of life, some organisms must therefore have undergone important evolutionary transitions that resulted in the kinds of species with which we are now familiar. In recent years, biologists have come to view these transitions not only as revolutions in the way living organisms looked and behaved, but also as solutions to similar problems. Understanding how they occurred brings us great insight into how natural selection works, and why modern, complex, organisms live and behave as they do.

Two people, John Maynard Smith and Eörs Szathmáry, did much to promote this conceptual unification in the 1990s (Szathmáry and Maynard Smith 1995; Maynard Smith and Szathmáry 1995, 1999). Together they defined eight major transitions (Figure 2.1), united by changes in the way that genetic information is transmitted between generations. In the origins of life they postulated individual replicating molecules forming populations of such molecules in compartments, such as cells (1). Later on these replicators bound physically together into **chromosomes** (2). Eventually, **RNA**, acting as both a replicator and **metabolic catalyst**, largely gave up these functions to more specialist molecules: **DNA** and **proteins** (3). Some **prokaryotes** (bacteria) eventually transformed into **eukaryotes** (4). Asexual clones among the eukaryotes transformed into sexual populations (5). Some single-celled **protists** transformed into multicellular organisms (plants, animals, and fungi) (6). In a few groups, solitary individuals began to live in social colonies (7), and in one of these, our own species, language emerged (8). We therefore bear the distinction of being the only lineage that has undergone all eight transitions. This would make us, in some quantifiable sense, the most complex biological entities not only in what we have evolved but in how we evolve.

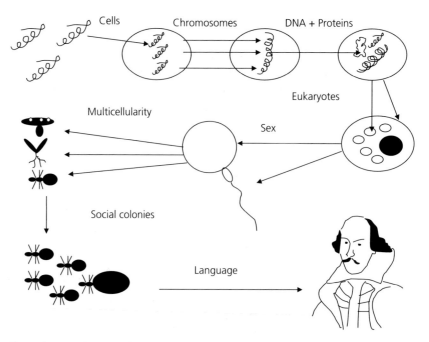

**Fig. 2.1**    The major transitions in evolution.

Explaining these individual transitions is challenging for three reasons, summed up by three different senses in which they are 'major'. The first, and the one that Maynard Smith and Szathmáry stress, is in an intellectual sense; the phenotypic changes we have to postulate are in themselves changes to the genetic system. This requires us to think especially hard about how evolution works because evolutionary biologists normally have the luxury of assuming that the genetic system is a constant. The second use of the word 'major' is in a structural sense: that the phenotypic changes were large. However, to be consistent with Darwinian evolution, changes must proceed by a series of small steps that retain functional integrity and which will be favoured by selection in each generation. We must first therefore imagine possible intermediate phenotypes, not all of which might be illustrated in the world about us. Then we must imagine environments or circumstances in which all the postulated intermediates would be favoured. In meeting these first two challenges we are postulating solely the origins of the characters involved. The third meaning of the word 'major', however, is that the changes were in some sense 'successful' from a macro-evolutionary perspective. In this sense we are implying that the transition was retained to the present day, and usually retained in abundance. This creates special challenges because, as will be shown later, the transitions can be seen to set up potential conflicts that would disrupt the integrity of the new system. Many of the transitions require formerly independent, or even totally new, genetic systems to come together and cooperate as part of a larger system. Yet, biologists are now used to the idea of genetic entities displaying selfish behaviour to ensure their own persistence. Thus it is sometimes problematic to imagine persistence of the novel unit. To cap off our problems, hypotheses must be consistent with existing evidence. Thin though that often is, even a little evidence can establish useful boundaries to possibilities, as fictional detectives are apt to explain.

In the third sense, the transitions were not initially major, but with the benefit of hindsight, having stood the test of time, many can now be seen to be so. To be retained in abundance, there are four possible contributing processes. First the transition might have happened on numerous occasions. In fact this is normally not the case. All of the transitions have happened to our knowledge only once, with the exception of multicellularity and social colonies, which have both evolved a limited number of times. This relative uniqueness is unsurprising given the drastic nature of the changes. The other three processes are; (1) reversal to the ancestral state, which might have been limited; (2) extinction of clades possessing the trait, which might have been reduced; and (3) speciation of clades possessing the trait, which might have been increased. The problem with explaining persistence is to find evidence for or against these processes.

## 2.1   Sex as a major transition

Let us see how one of the transitions stands up to these challenges. Sex technically refers to a special type of **cell cycle** (Figure 2.2), not, as is more normally used, copulation. Understanding the evolution of sex therefore means thinking hard about how cells replicate and divide and why this might change. Since the way cells do that is normally taken for granted, it is useful to be prepared for the unexpected in the paragraphs that follow.

Sex undoubtedly evolved in eukaryotes from a clonal ancestral state. A normal (**mitotic**) cell cycle is comparatively simple: some time into its life, each chromosome copies itself, and then the cell divides into two. In a sexual (**meiotic**) life cycle, new **diploid** offspring are born by fusion of two **haploid gametes** (syngamy). The gametes that fuse are normally very different in form and behaviour (anisogamy), one small and motile (sperm), the other large and immobile (egg). A number of mitotic cell cycles may then follow (development in multicellular organisms). Then each chromosome copies itself in a "premeiotic doubling" and form into homologous pairs each with two **chromatids**. These homologous pairs then swap bits of DNA (recombination), in a process known as 'crossing over' because of the appearance of the process under

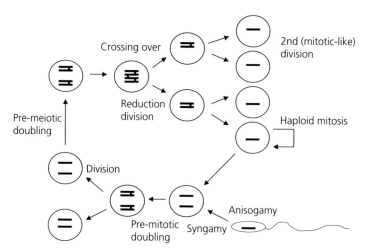

**Fig. 2.2**    A sexual life cycle. The dark lines are **chromatids** of a single homologous pair of chromosomes, drawn to indicate whether the cell is haploid or diploid, and whether the chromatids have replicated or not. The circles are cells.

the microscope. They then undergo two cell divisions to create four haploid cells. In some organisms these also undergo mitotic divisions before syngamy (Figure 2.2).

Which of these steps came first, according to Maynard Smith and Szathmáry (1995)? One of the surviving ancient protist lineages, *Barbulanympha*, which lives inside the guts of insects, has a cycle that involves endomitosis (gain of diploid state by copying of the haploid chromosomes) instead of syngamy. This led Cleveland (1947) to suggest that the first stage might have been the acquisition of a life cycle that alternated between a diploid stage acquired via endomitosis, and a haploid stage via a single one-step reduction division. Next, according to Maynard Smith and Szathmáry (1995), endomitosis would be replaced by syngamy. This would leave an otherwise normal one-step meiosis as seen in many sporozoans (the group to which the malaria parasite belongs). Crossing over and chromosome doubling followed, giving a two-step meiosis, and finally anisogamy (Figure 2.3). Let us see how we can account for one of those steps.

The vast majority of work on the evolution of sex has addressed the advantage of crossing over, or recombination. There are two processes that might have selected for its evolution. The first is that recombination can lower the **genetic load** if mutations act synergistically (having two is more than twice as bad as having one) (Kondrashov 1988). Imagine a distribution of deleterious mutations at equilibrium in a clonal population. Most organisms have a few, and a few have many (Figure 2.4). Now imagine that recombination occurs. The mutations are redistributed among the population, and once, after selection has acted and equilibrium is achieved, there are fewer mutations

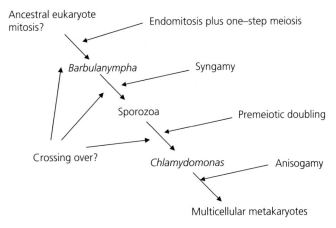

**Fig. 2.3**  Possible sequence of steps in the origin of sex, with intermediate states represented by some extant organisms, after Maynard Smith and Szathmáry (1995).

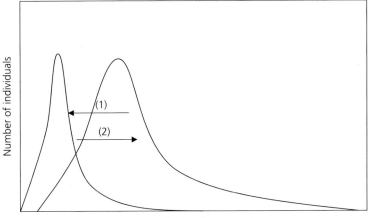

**Fig. 2.4**  The genetic load in sexual and asexual populations. Recombination can reduce the load of synergistic mutations (1). In small asexual populations, the genetic load of slightly deleterious mutations can increase, ratchet-like (2).

(Figure 2.4). This is because recombination in each generation throws together unlucky individuals with many such mutations, which suffer more severely than their fellows with only a few because of their synergistic effects. The death of these individuals purges the population somewhat of the mutations. This process can work even in an infinite population, and only requires the presence of synergistic mutations. There is presently little factual evidence for the synergistic effects of mutations, but theoretically it is a reasonable expectation (Szathmáry 1993). According to metabolic theory, mutations affecting a **metabolic cycle** should affect mainly the concentration of chemical intermediates, not that of end-products. If it is maximal end-product production that is important, as is likely in small organisms whose fitness depends on fast growth, then mutations are not synergistic. If it is some optimal balance of intermediates that is important, as in large long-lived organisms, then mutations will be synergistic and recombination will be favoured. Rather nicely, the frequency of recombination varies markedly across species, and is most frequent in large, long-lived organisms (Bell and Burt 1987), where we would most expect the effects of mutations to be synergistic.

The second possible process that might have favoured recombination is selection for change (directional selection) on polygenic traits (traits controlled at several loci) (Maynard Smith 1979). Hamilton (1980) most famously adhered to this hypothesis to explain not only the origin of sex but also more specifically its maintenance. He regarded **co-evolutionary** arms races (see Chapter 11) between hosts and parasites as a likely and widespread source of such directional selection. This idea has become colloquially known as the Red Queen theory (Van Valen 1973) after the Lewis Carol character in 'Through the Looking Glass' who had to run as fast as she could to stay in the same place.

The early protists were certainly not immune from such co-evolutionary forces, though the selective pressure is much greater on long-lived macroscopic organisms with long generation times relative to their parasites, where the traits under selection for change are those involved with defence and resistance. Nicely but at the same time frustratingly, this also fits well with the observation that long-lived organisms have higher rates of recombination. The frustration is that both Hamilton's and Kondroshov's hypotheses make the same prediction about the frequency of recombination relative to size and lifespan, so the observation fits but does nothing to narrow the range of plausible hypotheses. Rather more fortunately, there is independent evidence for the Red Queen hypothesis, which we will examine later in relation to the maintenance of sex.

Another step in the origin of sex is worth mentioning here. In many single celled organisms, such as the single celled alga, *Chlamydomonas reinhardii*, familiar in many school biology classrooms, the gametes that fuse to form a diploid alga are of identical size (isogamy). In most sexual species, however, one gamete of the pair (the egg) is larger and specialized to carry the **organelles**. Cosmides and Tooby (1981) and later Hurst and Hamilton (1992) argued that such specialization has evolved to prevent conflict between organelles from different parents. Many organelles, such as **mitochondria** and **chloroplasts**, contain their own DNA, (in the latter cases they were originally independent prokaryotic organisms). Such replicating entities should presumably be selected in the short term to produce copies of themselves at the expense of competing entities, and this is likely to be detrimental to the eukaryote cell as a whole. The potential problem is apparently real, for even in isogamous protists, uniparental inheritance of the organelles occurs, and in *Chlamydomonas* is apparently controlled by nuclear genes (central control, analogous to a police force in human society). Hurst and Hamilton argue that uniparental inheritance is possible only if there are two mating types and no more, for otherwise there is the danger of offspring lacking organelles entirely. Thus, conflict between organelles has, they claim, led to the origin of two (not three nor some other number) sexes. In fact, some protists exchange genetic information without **cytoplasmic** exchange (a process known as conjugation). In these cases there is no possibility of organelle competition, and multiple 'sex' or 'incompatability' types are known.

## 2.2   The maintenance of sex

Having drawn a scenario for the origins of sex, and thought of ways in which the various steps might be selected for, we are left with explaining its persistence and prevalence. It is clear that those lineages that evolved sex have gone on to diversify into millions of species that have largely retained sex. In some, however, clonal reproduction has secondarily arisen (these are normally called parthenogens). Is the commonness of sex due to the rarity of reversal to the clonal state? It is undoubtedly part of the answer. Some animals and plants, for example, have no known parthenogens, despite being species rich and well known. They include birds, mammals, and gymnosperms (conifers and their kin). In each of these three groups, mechanisms are known that are likely to have prevented reversal to the clonal state.

In birds, parthenogenetic individuals sometimes arise but these fail to persist as **unisexual** lines. The reason is that in birds the female is the heterogametic sex (with a 'Z' and a 'W' chromosome). During parthenogenesis, chromosome doubling occurs as normal, but there is only one subsequent division, leaving diploid eggs. Many of these will contain two Z chromosomes, leading to male production, and hence maintaining both sexes (Crews 1994). This is a big pity for short-term poultry production! If in birds, females were the homogametic sex (two X chromosomes), all parthenogenetic offspring would also have to be female. Given that in mammals females are the homogametic sex, one would imagine that cattle, sheep, and pig producers would have had more luck, but here prevention of parthenogenesis comes from another source.

In mammals the phenomenon that prevents parthenogenesis is known as genomic imprinting. The phenomenon was first noticed when researchers tried but failed to get embryos to develop by fusion of two egg nuclei. The reason is that early zygote development requires genes from both parents that have different levels of activation: if the genes come from the same parent the activation levels are all wrong. We will encounter this phenomenon again in Chapter 7, but, briefly, the reason is likely to be that in mammals parents conflict over what level of gene activation in the zygote is preferred, setting up an offensive/defensive gene activation war (e.g. Burt and Trivers 1998).

In gymnosperms the mystery is more clear-cut: although the egg provides most of the organelles for the **zygote**, it is the pollen that provides the chloroplasts. Unisexual gymnosperms would lack the ability to **photosynthesize**. Provision of an essential organelle is also the likely reason for general rarity of animal parthenogenesis, and also for the strange forms it takes when present. In many animals, the sperm provides a **centriole** to the zygote. In many parthenogenetic animals, including most clonal vertebrates, parthenogenesis is 'sperm-dependent': the eggs need to be 'fertilized' by sperm of another species for successful development, though the sperm **genome** never makes it into the next generation (Beukeboom and Vrijenhoek 1998).

Though obstacles to reversal are one important reason why sex is still prevalent, it is not the whole story. Many plants, for example, could easily persist from generation to generation in a clonal state by vegetative reproduction. Yet, where they exist, wholly clonal plants, and parthenogenetic organisms in general, appear to be relatively recent phenomena. For example, most are isolated species in genera that are predominantly sexual. In a few cases, genetic and other techniques have been used to estimate the actual ages of clones and their sexual parents. In the case of the genus *Poeciliopsis*, a guppy fish that inhabits streams in southern United States and Mexico, sexual species are up to 3 My old, but their sperm-dependent parthenogens are mostly less than a few thousand (Figure 2.5). Interestingly, the oldest surviving clone is also the only one where the sperm genome finds

**Fig. 2.5**   The unisexual fish *Poeciliopsis 2monacha-lucida*. This species is a sperm-dependent parthenogen, meaning it relies on the sperm of another species for reproduction. However, the genome of the sperm donor is not expressed in the offspring which are genetically identical to their mother. This species suffers more from parasites than its sexual relatives, hence supports the 'Red Queen' hypothesis for the maintenance of sex. Photo courtesy of Bob Vrijenhoek.

expression in the fish phenotype, the so called 'hybridogen' (it is later excluded from gamete production though). Only recent originations persist probably because earlier originations have largely gone extinct while their sexual relatives have not (Vrijenhoek 1994). There are a relatively few higher taxa that have persisted for millions of years, popularly known as the 'ancient asexuals'. Our theories for why parthenogens might have high extinction rates should ideally account for these exceptions.

The problem of the relative persistence of sexual species versus asexual clones has traditionally been thought of in terms of the so-called 'two-fold cost' of sex. Female parthenogens can increase in number at twice the rate of their sexual counterparts because they do not give birth to males, which cannot themselves give birth (Figure 2.6). Sexual organisms may even suffer a number of additional costs, such as finding a mate. All of this suggests that sexual organisms should be the ones with the higher extinction rates. It was this problem that eventually became the focus of Hamilton's research: why was it that clones, once they arose, did not quickly send their sexual parents extinct? Hamilton became convinced that the answer lay with one of the short-term advantages of recombination, in particular the Red Queen hypothesis. There is evidence to support his contention. First, some

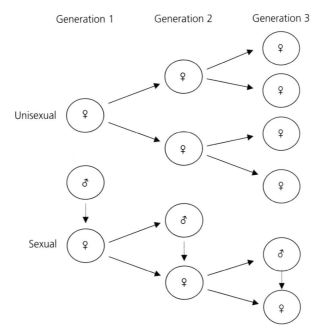

**Fig. 2.6**   The two-fold cost of sex. Here a species is shown which always has two offspring per generation, except that in the sexual form half of these are male, which mate with the females (dotted arrows), while the unisexual form can have offspring without mating. By the third generation, there are four females in the unisexual line but only one female in the sexual line.

of the so-called ancient asexuals are obligate mutualists. These include some mycorrhizal fungi, which inhabit the roots of plants, and the fungi that are the food for leaf-cutter ants. In contrast to parasites, which experience **directional selection** from their hosts, mutualists would be expected to experience **stabilizing selection** to aid the efficiency of the interaction. This, as Maynard Smith showed, can select for an absence of recombination. In addition, we have direct measures from some asexual clones that they carry higher parasite loads than their sexual counterparts. This is true, for example, of the *Poeciliopsis* clones (Vrijenhoek 1994, Figure 2.5).

Clones, of course, may also suffer a higher load of deleterious mutations, as Kondroshov showed, contributing to their extinction rate. In *Poeciliopsis*, there is also evidence for this. By a clever series of crosses, it has been possible to express the hybridogen genes that are normally dominated by those of the sperm donor. These show several developmental defects compared with the parental or hybrid genotypes.

In addition to Kondrashov's mechanism, an alternative long-term mechanism can account for this, known as Müller's ratchet. Müller's ratchet is in some ways more general than Kondrashov's theory, for it does not rely on synergistic mutations; mutations merely have to be of small effect. It also only works in small populations, but that is probably a fairly general phenomenon. When mutations are of small effect, the distribution of those mutations at equilibrium among individuals will be approximately bell-shaped: few will be completely free of them, most will have a few, and only a few will have a lot (Figure 2.4). In a small population, however, the categories of individual with no mutations is easily lost by chance events, even if they are the most fit. In a clonal population, these can never be recovered. Of course, the category of individuals with most mutations can also be lost, but those are replaced by subsequent mutation.

Overall then, in a clonal population, the load of slightly deleterious mutations continually cranks up, ratchet-like. In a sexual population, however, recombination recreates individuals free of mutation, and the genetic load remains stable despite stochastic loss of the fittest individuals (Figure 2.4). Note that because of the long-term nature of the mechanism, it cannot be invoked to explain the origination of recombination, merely its persistence relative to clonal reproduction.

In general, should we be searching for long- or short-term mechanisms to explain the persistence of sex? Hamilton was convinced that the latter was necessary largely because of competition between clone and sexual parent. Is there in fact evidence for this? Do clones actually displace their sexual parents, and is there a risk of them being sent entirely extinct? In general, we might expect, in the absence of short-term advantage, that the clone would successfully displace the parent from part of its former range, but that existing genetic diversity among the parent would allow the parent to persist in parts of its range to which the clone is less adapted. In *Poeciliopsis*, the diversity of species is inversely related to the diversity of clones, suggesting some competitive exclusion. In many plants, such as the dandelions, parthenogenetic forms also appear prevalent in some environments, especially at high altitudes and latitudes. This in turn suggests that the costs and benefits of each form of reproduction vary spatially, and also that short-term advantages of sex are not always initially enough to compensate for any costs.

To summarize, the maintenance of sex has been influenced by the following processes: first, constraints to reversal among a number of lineages, particularly animals. Second, clones can successfully displace their sexual competitors from some environments, suggesting a severe cost to sex. However, directional selective forces can compensate for these costs in some environments, and, combined with increased genetic loads, send most clones extinct relatively rapidly.

How does the evolution of sex compare with the other major transitions? Maynard Smith and Szathmáry identify several common features of the transitions (Table 2.1). Of these, sex is very illustrative. Entities have combined together, through the evolution of syngamy, to form a sexual population from previously independent clones. Further, reversal of sex to the clonal state is sometimes difficult because sex has developed a complex machinery for reproduction. Sex has probably led to conflict between entities, such as between organelles in the parent gametes for representation

**Table 2.1** The common features of the major transitions in evolution

| Feature | Molecules in compartments | Chromosomes | DNA + protein | Eukaryotes | Sex | Multicellular life | Social colonies | Language |
|---|---|---|---|---|---|---|---|---|
| Entities combine? | Yes | Yes | No | Yes | Yes | Yes | Yes | No |
| New ways of information transmission? | No | No | Yes | No | No | Yes | No | Yes |
| Reversal difficult? | Yes | Yes | Yes | Yes | Yes | Yes | Yes | ? |
| Conflict between entities? | Yes | Yes | No | Yes | Yes | Yes | Yes | No |
| Mechanisms to prevent conflict? | Yes | Yes | No | Yes | Yes | Yes | Yes | No |
| Division of labour? | Yes | Yes | Yes | Yes | Yes | Yes | Yes | No |

in the zygote. To solve this, mechanisms, such as uniparental inheritance have evolved. Sex has led to division of labour among the combining entities, such as male and female gametes. In the evolution of sex, there has been no new method of transmitting information developed. That has only occurred in three of the transitions: in the origin of DNA and protein from RNA, in the origin of language, and in the origin of epigenesis (gene activation) in the origin of multicellular life.

In this chapter we have been postulating processes that have caused change within lineages through natural selection, a theme that will continue in the next several chapters. I cannot help but end here with a well-known quote from Aldous Huxley who once said that 'an intellectual is a person who has discovered something more interesting than sex'. There is a certain irony in this quote for evolutionary ecologists. Many of them would argue, on purely intellectual grounds, that there is in fact nothing more interesting than sex, full stop.

## 2.3   Further reading

Beginners should try Maynard Smith and Szathmáry (1999) first, followed by Szathmáry and Maynard Smith (1995). Maynard Smith and Szathmáry (1995) is quite heavy going, but also more complete. Useful works on the evolution of sex include Maynard Smith (1984), Bell (1982), Stearns (1987). Two recent special issues cover the subject: in *Science* (25 September 1998, vol. 281: 1979–2008) and *Trends in Ecology and Evolution* 1996 vol. 11. *Poeciliopsis* is reviewed by Vrijenhoek (1994).

# 3 Brave new worlds

The eight novel ways of transmitting information, outlined in the previous chapter as 'major transitions', by no means exhaust evolution's extraordinary feats. Over the history of life, evolutionary events have also radically changed the characteristics of the **biosphere**. As in the previous chapter then, which considered how evolutionary changes increased the complexity of organisms, this chapter will consider how evolution has increased the complexity of planetary ecology. Four things generally indicate that major ecological changes have occurred in the past, identifiable, for example, from the fossil record (Vermeij 1995; Kanygin 2001): First, changes in species richness. Second, organisms living in new places. Third, ecosystems with new **functional groups**. Fourth, new **geochemical cycles**. For present purposes we are only interested in such changes that are linked with evolutionary events, as most (but not all) of them are by definition. The changes need collectively to create the essential and complex features of modern planetary ecology, which are worth briefly describing.

Species richness is currently (recent extinctions excepted) higher than ever before (Signor 1990; Sepkoski 1999). About half of described **macroscopic** species are insects, a quarter green plants, and most of the rest sundry invertebrates (Southwood 1978). Life exists and flourishes nearly everywhere on the face of the globe; in the marine, freshwater and terrestrial realms, anoxic mud, animal guts, hypersaline lakes, volcanic springs, desert dunes, within frozen antarctic rock and in rocks hundreds of metres below ground. Some organisms even live an essentially atmospheric existence, feeding on other flying organisms.

The most productive and diverse ecosystems on the planet are terrestrial. They comprise: green plant producers; a diverse assemblage of vertebrate and insect herbivores; predators, parasites and **parasitoids**, and a soil fauna of scavengers, **detritivores**, decomposers, and nutrient cycling bacteria. In the marine realm the most productive and species-rich ecosystems are those with sessile producers, such as coral reefs, kelp forests, and seagrass beds. The open ocean is a comparative desert, but consists of pelagic **phytoplankton**, **zooplankton**, and macroscopic predators. A **benthic** fauna consists of filter feeders, scavengers, and predators. The biosphere's energy comes, almost exclusively, from light captured by plants, and its carbon source is atmospheric

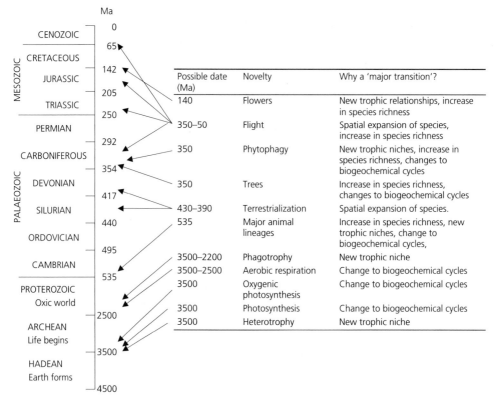

**Fig. 3.1**  Some major transitions in ecology resulting from evolutionary novelty, and when they occurred.

carbon dioxide, which is eventually returned by (high energy-yielding) aerobic respiration. Where, then, did all this come from and what evolutionary novelties helped it get there? Below I tentatively identify eleven major changes that take us from the origin of life to an ecologically modern planet (Figure 3.1).

## 3.1    Evolution of the biosphere: a brief history

The origin of the biosphere and of earth's ecology occurred between 3.8 and 3.5 billion years ago. Both **autotrophic** and **heterotrophic** origins have been proposed. Previously the heterotrophic use of organic molecules synthesized in the pre-biotic broth was a popular idea (see Lazcano and Miller 1999). More recently, autotrophic theories have re-emerged, for two reasons: first, early **cell membranes** would probably have lacked sufficient permeability to transport large molecules. Second, it is now realized that metabolic cycles that grow and reinforce themselves can emerge spontaneously. If such metabolic cycles occurred in the pre-biotic world, an autotrophic ancestral metabolism would by definition result (Wächtershauser 1988, 1990). Because photosynthesis today requires a more complex biochemistry, an ecosystem consisting initially of only **chemo-autotrophs** is likely. It is also likely that a second trophic level would quickly be added, consuming waste products (and dead remains) of these producers. There is evidence also that photosynthesis may nonetheless have been a very early acquisition. Early life probably lacked protein **enzymes** to catalyse reactions, using instead RNA. It turns out that **chlorophyll** synthesis involves (non-intuitively) molecules bound to RNA, a likely relic of the ancient involvement of RNA in catalysis (Benner *et al.* 1989). The origin of photosynthesis provided a new energy source, light, for the world's ecosystems that would have massively increased its potential productivity.

The early photosynthesizers probably used hydrogen sulphide or molecular hydrogen as their source of electrons, rather than water (Xiong *et al.* 2000). Using water as an electron source releases molecular oxygen. We know that **cyanobacteria**, which today are the predominant oxygenic prokaryotes, were among the earliest prokaryotes, at least 3.5 billion years old. However, it was not until about 2.5–2 billion years ago that a great increase in atmospheric oxygen occurred, marking the end of the Archean era (Figure 3.1). Until then, some process must have removed the oxygen produced by cyanobacteria. One possible process is aerobic respiration (Towe 1990). Aerobic respiration is many times more energy-yielding than any of its anaerobic alternatives, and although at this stage there were certainly no food chains as we recognize them today, aerobic respiration made possible the future advent of the higher trophic levels with substantial biomass (Fenchel and Finlay 1995). By this very early time, 3.5 billion years ago, all basic bioenergetic processes had probably evolved, many of them several times, and the biogeochemical cycling of carbon, nitrogen, and sulphur was established as we know it today.

Although the earliest recognizable eukaryote fossils date from the time of transition to oxic atmosphere, 2100 Ma, the lineage from which modern eukaryotes are derived is a very deep evolutionary branch, and probably dates right back to the time of the earliest evidence for life, 3.5 Ga. Somewhere in this interval, one of our ancestral bacterial lineages (biomolecular evidence suggests it resembled an **archaebacterium**) developed a **cytoskeleton** and lost its **cell wall**. With this came the ability to engulf large organic particles. The first eukaryote lineages were doubtless anaerobic, a fact indicated by the many **extant** anaerobic eukayotes belonging to basal branches of the eukaryotic tree. These early predators represented a trophic interaction quite different to anything in the prokaryotic world, in which organic material had to pass through cell walls and membranes. However, only with the advent of mitochondria, probably at the Proterozoic boundary

(Figure 3.1), could these predators fully reap photosynthetic productivity by integrating with the aerobic world.

There existed now a period of about 1 Ga that is relatively featureless in terms of historical evidence, and indeed may have been relatively stable in ecological terms, until 535 Ma. The next 100 My period, known as the Cambrian explosion and subsequent Ordovican radiation, is one of enduring interest for biologists (Figure 3.1). Over the next 100 My appeared the major animal lineages, including the first benthic and pelagic macropredators and the first animals capable of burrowing more than a few millimetres into sediments.

The consequences were various and significant. The Earth became much more species rich. Use of skeletal materials based on calcium, phosphorus, and silica led to greater control of these minerals by organisms, as opposed to by inorganic processes. Disturbance of sediments by burrowers recovered carbon and other nutrients from sediments for recycling rather than burial. All these new animals produced masses of faeces. Faeces dropped to the ocean floor rather than remaining in the water column, and in doing so consumed less dissolved oxygen. Flow of oxygen from the surface waters to the ocean floor would have been facilitated, as suggested by geological evidence (Logan *et al.* 1995). This may itself have contributed to the Cambrian radiation by facilitating skeletal formation, large bodies, and active metabolisms. Macropredators might have stimulated novel defences in prey, which themselves would be a cause of selection on predators. Such 'co-evolution' may have been a stimulus for diversification in lifestyle and structure.

By the end Ordovician, marine ecosystems would have looked pretty modern. Soon after, however, multicellular organisms then began to form complex terrestrial ecosystems. At the time, colonization of the land was notable primarily for an expansion of earth ecospace. However, eventually, the progressive evolution of terrestrial communities led to major alterations of biogeochemical cycles, and terrestrial domination of global biodiversity and production. The earliest land plants were relatively small in stature. About 380 Ma the earliest trees appeared and by 350 Ma, forests composed of horsetails, clubmosses, ferns, progymnosperms, and seed plants had a widespread global distribution and covered a number of clearly distinguishable biomes. The consequences for the biosphere appear to have been immense. Global productivity probably soured to unprecedented levels. Coal was deposited in massive amounts never again attained. Global carbon dioxide levels dropped to 10% of their previous levels in about 50 My, eventually resting about their present level. This set the scene for subsequent periods of significant global cooling.

It is possible that the earliest terrestrial plants were relatively free from natural enemies, such as herbivores. By mid-Carboniferous there was abundant evidence that the onslaught had begun. Insects with characteristic mouthparts, and fossil leaves with evidence of bite marks, along with vertebrates with teeth designed for chewing all suggest that plants had begun their war against animal attack that continues today. This was a significant new trophic level, for now some of the plant productivity was available to other organisms. Terrestrial ecosystems had come of age.

Flight has probably evolved four times in the history of life: in insects, pterosaurs, birds, and bats. Most biologists would agree that flight has had major ecological repercussions. First, the atmosphere could at last be properly utilized. While flying species are variably adapted to an aerial existence, a few birds, such as the swifts, live the vast majority of their (often considerable) lives on the wing. Flight may also have contributed significantly to global diversity. Bats, birds, and winged insects are all species rich (see de Queiroz 1998). These organisms are likely to have contributed to diversification of other species, such as the plants they pollinate and disperse. The evolution of flowering plants is our final major transition. Angiosperms are the most species-rich division of plants today, they dominate global productivity, and their origins coincided with a rise in plant diversification that shows no signs of abating. Effects on insect diversification are also detectable and non-trivial (Farrell 1998). We have now arrived at an essentially modern ecology. How and why did these changes occur and why have they been retained? Let's look at an example, the evolution of flight in birds, insects, bats, and pterosaurs.

## 3.2 The evolution of animal flight: understanding a major transition in ecology

There is now little doubt among most biologists that birds derive from a group of theropod dinosaurs. The theropods were a bipedal carnivorous group that share many anatomical features with birds. A series of recent fossils, most notably from Liaoning province of China and described by Xu Xing and colleagues, include theropods with epidermal feather-like structures that we might collectively refer to as 'fuzz'. They were pre-adapted for flight through a fast **cursorial** predatory lifestyle. This resulted in a shortening and stiffening of the tail, reduction in the size of the midbody, lengthening of the **raptorial** arms, swivel-wrist joint, light hollow bones, and a reduction in body size (Sereno 1999).

What subsequently happened? There are several ecological scenarios. The arboreal hypothesis states that birds evolved from ancestors that lived in trees and gained the ability to glide from tree to tree. Arboreal gliding organisms are common today, and in general provide a plausible intermediate stage to flight because the energy for lift is supplied by gravity. The discovery of *Microraptor ghui*, with its apparently four gliding limbs (Xu *et al.* 2003), has recently renewed interest in this scenario. However, theropods were primarily bipedal ground dwelling runners and this has also focussed attention on a possible cursorial origin. By flapping their forearms as they ran, rather like a swan taking-off from water, theropods could have increased their running speed by taking weight off the hind legs, allowing the hind legs to provide more forward thrust. In this way the wings would gradually take over from the hind legs until both lift and forward thrust could be provided by the wings alone (Burgers and Chiappe 1999).

Other more complex scenarios have been also proposed. Garner and co-workers (1999) have suggested the 'Pouncing *Proavis*' hypothesis. They envisage theropods dropping onto prey from a perch, in the manner of modern owls and buzzards. The forearms could have assisted balance and the feather surface would develop initially to control the drop, rather than provide lift. Selection for greater horizontal range of drop (a swoop, involving lift generation) could then have transformed the role of the wing. Like the cursorial hypothesis this retains the functional distinctiveness of the fore- and hindlegs since both have separate roles, which is less likely in an arboreal origin. Another possible scenario (Dial 2003) is that wing flapping developed to assist theropods to climb to elevated refuges, such as trees or bolders, as it is still used today in birds, such as quails and chickens, even in their flightless chicks.

If birds developed flight from bipedal and basically ground-dwelling ancestors, what of insects? There is intriguing evidence that insect wings may have developed from leg segments supporting gills that originally developed along the length of the body. Some fossil mayfly nymphs possessed these, and the thoracic ones would then be homologous with modern wings (Kukalova-Peck 1978). In modern insects, genes are present that switch off the development of these structures in all but the wing segments, but the potential to form wings is present in every segment (Carroll *et al.* 1995). But how could an aquatic gill come to function as a wing? Some mayflies and stoneflies use their wings in a unique way (Figure 3.2): to sail or skim across the water surface to reach land after emerging as adults onto the water surface (Marden and Kramer 1995). Close to the water's surface, wing beating provides more power because the air is compressed between the wing and the water. The aquatic skimming origin therefore postulates ancestral **apterygotes** and **pterygotes** having aquatic larvae with moveable gills, and air-breathing adults. Retaining gills through to the adult stage aided sailing to land, and selected for larger but also fewer structures to aid directional stability. Skimming developed further speed and control via flapping, and eventually

**Fig. 3.2**    This male stonefly, *Allocapnia vivipara*, is flightless but has raised its short wings to sail across the water surface to dry land. Flight in insects may have originally evolved through such a stage. Photo courtesy of Jim Marden.

adults were created that were fully capable of flight. The hypothesis provides an explanation for the somewhat mysterious observation that while the primitive wingless insects and most derived insects are terrestrial, the extant primitive winged insects (mayflies and dragonflies) all retain aquatic larvae.

The origins of bats and pterosaurs are much more obscure. Neither have clearly identifiable fossil ancestors, nor do transitional fossil forms exist. In both taxa the flight membrane and lack of cursorial hind legs are much more suggestive of an arboreal gliding ancestor than for the birds and insects (Figure 3.3). Some candidate pterosaur ancestors may have been bipedal however, and bipedality was certainly common among the stem **archosaur** groups. In the bats, most scenarios envisage a nocturnal, arboreal, and insectivorous ancestor for the following reasons: the hind limbs help support the flight membrane (Figure 3.3), making a cursorial ancestor very unlikely; all bats are nocturnal hence that is a likely ancestral state; and the ancestral **eutherians** were doubtless insectivorous. Recently, Speakman (2001) has proposed an alternative hypothesis: that of an arboreal, diurnal, frugivorous ancestor. The advantage of this hypothesis is that one can imagine the ancestral bat leaping from branch to branch using vision effectively for foraging. Some then developed insectivory, and all were forced into nocturnality to escape raptorial birds, after which fruit bats specialized the visual system, and the microbats the echolocation system. There is agreement that the setting was arboreal and gliding, but beyond this many scenarios are possible. The origin arguments presented above are summarized in Table 3.1. It is exciting that of the three extant flying taxa, three completely different evolutionary scenarios *may* have played out.

**Fig. 3.3**    A lesser mouse-eared bat, *Myotis blythii*, taking to the air. Note the short hind limbs and flight membrane stretching from the forelimbs to the hind limbs. This is very un-bird-like and reminiscent of a quadrupedal glider. No bats have become flightless. Photo courtesy of John Altringham.

So we can imagine plausible scenarios for how these origins may have happened. Why to these organisms though, and why at those moments in time? There are dozens of gliding animals in today's forests: why have they not all developed powered flight? Have they simply not hit on the necessary mutations to transform a gliding animal into one with powered flight? It is unlikely. Our evolutionary understanding of the transitions involved is that they have been of a continuous nature, with small change building upon small change. This is well illustrated by the fossil record for bird evolution. None of the changes involved can be seen as particularly remarkable. Furthermore, birds, insects, and bats all have a different flight apparatus suggesting that selection can work through multiple routes. More likely then, specific external influences may be necessary. Recently, Dudley (2000) has championed the view that increases in atmospheric oxygen concentration may have facilitated the origins of powered flight. The Late Carboniferous period, characterized by the first pterygote insect fossils, represented the historical peak in Earth's oxygen concentration, about double that of today. The Mesozoic era, characterizing the origins of pterosaurs and birds, was a period of increasing oxygen concentration, reaching a secondary peak in the Late Cretaceous, reducing somewhat since. This secondary peak coincides

**Table 3.1** Hypotheses on the origins of flight

| Group | Who changed? | What changed? | Where did it change? | Why did it change? | Controversies |
|---|---|---|---|---|---|
| Pterygote insects | Apterygotes with aquatic larvae | Larval gills derived from pleural structures became wing | Newly emerged adults sailing or skimming over the water surface | Improved speed over the water surface | No consensus on the origins of wing structures, or on aquatic apterygote ancestors |
| Pterosaurs | Unknown basal archosaur | Gliding membrane supported by a finger became wing | Arboreal setting | Improved distance and control of glide | Absence of any fossil ancestors makes this an open question |
| Birds | Feathered dromeosaurs | Feathered forearm became a wing | Cursorial setting | Flapping aids running speed by providing lift | Primarily about the ecological setting and reasons for change |
| Bats | Unknown basal eutherian | Gliding membrane supported by fingers became wing | Arboreal setting | Improved distance and control of glide | Absence of any fossil ancestors makes this an open question |

with the origins of flight in vertebrates. High oxygen concentrations would have two important effects: first, they would have increased the density of the air, and the flight surface would thus provide more lift. Second, they would increase metabolic capacity and hence provide more power per effort. Since flight is energetically expensive, slight increases in power and lift might have been sufficient to turn net cost into net benefit. Rather interestingly, changes in oxygen concentration may have been stimulated by other evolutionary transitions, such as the increase in productivity due to terrestrialization. Recently, Lenton *et al.* (2004) have suggested that a chain of such events might have occurred in the history of life, with evolutionary changes giving rise to environmental changes, which in turn give rise to evolutionary changes.

## 3.3 The maintenance and ecological effects of flight

Once flight originated, for its ecological effects to be expressed it had to be maintained. Loss of flight is an interesting phenomenon, for it has occurred very frequently in insects and birds, and not at all in bats (nor probably in pterosaurs). The obvious difference between these two groups of organisms is that the former (insects and birds) retained a functional distinction between the flight apparatus and the legs: they can both walk or run without using their wings. In pterosaurs and bats this is not the case and both groups would be relatively ineffective on the ground, hindering the transition to a terrestrial lifestyle again. In birds and insects, loss of flight could mean a reallocation of energy away from the flight apparatus and increases in reproductive expenditure (Roff 1990, 1994). They lost much of course, and in birds loss of flight is only viable under special circumstances. It has happened mostly in a few taxonomic groups (rails notably) and under special ecological circumstances (notably on islands, see Figure 13.3) (Roff 1994). There are three likely reasons for the latter: first, an absence of land predators that makes escape (and especially nesting off the ground) less important. Second, on islands, high dispersal tendencies might increase the risk of mortality through loss of individuals at sea. Third, a less active metabolism might be very advantageous in surviving long periods of food shortage on islands, where birds cannot simply move elsewhere.

Insects have lost their powers of flight many more times than birds, and they have done so in a variety of ecological circumstances (at least once in nearly all major habitats). There are many flightless island insects, no doubt many for the same reasons as birds, but there are also many flightless insects in other habitats. Loss of flight is very rare in freshwater insects. This is unsurprising given the ephemeral nature of many freshwater habitats, most of which stand a good chance of temporary or permanent drought. Conversely, flightlessness is very common among parasitic insects, particularly of vertebrates (Figure 3.4). In fact only two large radiations of secondarily flightless insects have occurred: among the fleas and the lice, which primarily use mammals and birds as hosts. These do not require flight for dispersal to new hosts. Wing loss may also bring a particular advantage in facilitating movement within the fur and feathers of their hosts. Given that loss of flight is apparently so easy in insects, one may wonder why only two large radiations of flightless insects have occurred. One possibility is that speciation is frequently associated with niche shifts that would favour the reversal of winglessness again. Another part of the answer may be that, in general, flightlessness leads to poorer net rates of cladogenesis. Both extinction risk and speciation probability may be affected and both these subjects will be the focus of later chapters.

The maintenance of flight enabled it to become a major transition. Why did this transition have ecological repercussions? I have argued above that these were manifested in three main ways: first, flight in itself represents an occupation of ecospace (the atmosphere) previously unoccupied, just as terrestrialization does. This requires no special explanation: the innovation

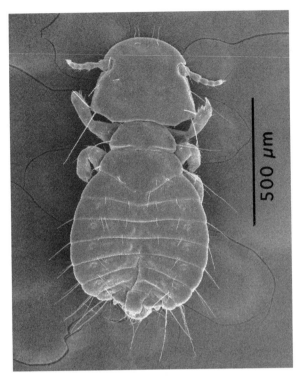

**Fig. 3.4**     A body louse, *Campanulotes compar*, 1 mm long, which infests feral pigeons, *Columba livia*. Lice and fleas represent some of the few major radiations of insects that have lost their wings. Photo courtesy of Sarah Bush and E. King.

and the ecological change are synonymous. Second, flight probably also increased the species richness of clades that evolved it (de Queiroz 1998). It is interesting to speculate why this might have been. Dispersal is often linked with the speciation process because it can transport organisms to isolated areas where they can differentiate from their ancestors (Chapter 1). Dispersal may also inhibit extinction by allowing areas in which local extinction has occurred to be recolonized. Flight may also have had interesting repercussions on life history (see Chapter 4). Birds, for example, have considerably extended lifespans compared with mammals of the same mass. Bats also have much longer lifespans than other mammals of the same mass. It does not necessarily follow that an increase in lifespan will reduce the extinction risk of species, but it is possible. This theme is investigated in more detail in Chapter 14. Of course, reduction in mortality, such as from predators, may have helped select for powered flight in the first place. Third, flying organisms perform important ecological roles that may have stimulated the evolution of other groups, the most obvious case being plant–pollinator interactions. Families of angiosperm with animal pollination are significantly more species rich than those without animal pollination (Ricklefs and Renner 2000). Insects, bats, and birds are the major pollinating animals and they can also fly. Their dispersal abilities give them much greater potential to visit other flowers and promote outcrossing.

The evolution of flight then, illustrates well the challenges, but also the rewards, of understanding a major transition in ecology. We have to consider origins, and in particular, scenarios of progressive selective advantage in the face of small continuous changes. Those are available for birds and insects, and to a lesser extent for bats and pterosaurs, and the scenarios are diverse. We must also explain the timing of the events and what stimulated them. Invoking high metabolic requirements for flight and periods of high oxygen concentration may explain that, implying that changes in the environment were critical, perhaps stimulated by previous evolutionary novelties. We also have to consider the frequency of loss of the character and its relative effects on speciation and extinction rates. Losses of the character are few in some groups, where the flight mechanism is at odds with terrestrial locomotion, and common in others where it is not. Where it is common, losses do not seem to confer any consistent benefit in terms of increased speciation or reduced extinction rates. Finally, we need to explain why the ecological effects of the transition have been great. Some of these themes may also be found in some of the other transitions, and it will be exciting to discover if a consensus on this emerges in the coming years of the sort that Maynard Smith and Szathmáry provided in the previous chapter.

## 3.4 Further reading

Southwood (2003) provides an introduction to the history of evolutionary and ecological change. Fenchel and Finlay (1995) and Willis and McElwain (2002) are useful more specialist sources. Padian and Chiappe (1998) give a good introduction to bird origins, Thomas and Norberg (1996) summarize the insect story, and Speakman (2001) bats. Dudley (2000) deals with the when. Roff (1990) covers many issues of flightlessness.

# 4 Traits, invariants, and theories of everything

The last two chapters were concerned with how natural selection can change the way a lineage looks (its phenotype). These changes have had profound effects, first on the complexity of organisms (Chapter 2), and second on the complexity of the ecology of the planet (Chapter 3). This chapter continues the theme, but deals with traits that are individually far more ordinary: life history traits. Nonetheless, they account for much of the diversity in form across species, thus collectively are of immense interest.

A life history is a description of the major characteristics of an organism from its birth to its death. The traits are variables that can be measured or categorized across individuals and that collectively make up the description (Figure 4.1). In the past, most work on life history evolution proceeded on a trait-by-trait basis. Each of the traits is associated with a large body of literature asking the following question: what selective forces have driven the

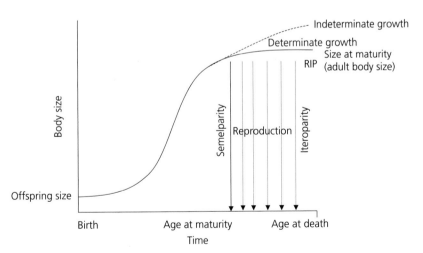

**Fig. 4.1**   A cartoon of a life history, showing size of an individual against time, and various life history traits highlighted. Semelparity means reproducing only once, iteroparity many times. Determinate growth is when growth stops at reproductive maturity, indeterminate growth when it continues.

trait into its present states? Typically, theoretical models are constructed for each trait, describing the kinds of environment that favour different values of the trait. The models are then tested by seeing whether their assumptions and predictions are met in nature. If the answer is yes in both cases, our model's assumptions provide us with the understanding we are looking for. If the answer is no, either the model or the data are inadequate and must be re-examined.

In general we might expect certain values of a trait to be favoured rather universally. For example, it is clearly best, in a Darwinian sense, to produce many offspring, all other things being equal. Happily for those who like diversity, all other things are not equal: high fecundity must come at a cost to some other trait of importance to the organism's fitness, such as offspring size. This phenomenon is known as a trade-off. The organism is thus faced with the more complex choice of finding the value of the trait that maximizes fitness in the face of trade-offs with other traits of importance. In the above example, evolution would essentially select between organisms that have large offspring but few of them, and organisms that have many but small offspring. Following this, our evolutionary model will generally make some assumptions about the nature of any trade-offs involved (a constraint), and about the consequences of each value of the trait for fitness (a currency). In general, it is assumed that the value of the trait (the strategy) that leads to greatest fitness in a given environment will be that which is selected for, and should, all being well, represent that observed in nature (Figure 4.2). Such models are commonplace in evolutionary ecology and are known as

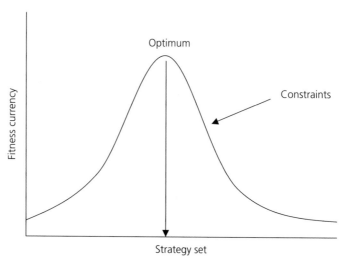

**Fig. 4.2**    The characteristics of an optimization model. The arrow shows peak predicted fitness and therefore what would be expected in nature.

optimization or optimality models (Parker and Maynard Smith 1990). Let us start by seeing how this approach works.

# 4.1 Single trait optimization: reproductive lifespan

Some organisms display 'big bang' reproduction, after which they die (semelparous), while others reproduce more consistently over a long period of time (iteroparous) (Figure 4.1). The question of whether to be semelparous or iteroparous is a rather simple one because there are only two strategies to consider. We simply have to calculate, for any given organism, which should lead to the highest fitness. Charnov and Schaffer (1973) produced a simple calculation to help, which is derived by asking which strategy produces the highest rate of population growth (their fitness currency). They assumed a plant could display either an annual or a perennial strategy. The fitness of the annual is determined by its own survival to maturity and fecundity once mature. However, that of a perennial is determined by both its juvenile survival, annual fecundity, and adult survival from year-to-year. These are the constraints. If annuals are to be more fit than perennials, annuals must have higher annual fecundities than perennials, because they do not survive for more than one flowering season. In contrast, high adult survival should select for perennials because this allows them to increase population growth rate, the fitness currency.

The data provide confirmation of these predictions. Annual plants generally have higher annual fecundities than their perennial close relatives (generally between 1.5 and 5 times) (Young 1990). Such fecundity differences are most apparent between close relatives inhabiting adjacent environments. On Mount Kenya in Africa, for example, are two species of *Lobelia*, one semelparous and the other iteroparous (Figure 4.3). The semelparous forms have high fecundity and grow on dry rocky slopes, where adult mortality is very high. Iteroparous forms have lower annual fecundity and grow on moister valley bottoms where adult survival is higher (Young 1990).

The development of simple (classical) models like the above has been repeated for all of the common traits under investigation. But a 'single trait' view of the world is also somewhat limiting, for each organism represents a specific and sometimes characteristic combination of traits. To understand the organism as a whole we would have to apply and test theory for each of the life history traits separately. This is clearly an impractical task on a large scale. Instead, we need an approach that considers, in a single framework, the whole life history of an organism. That has been the recent challenge in life history evolution, and it has led to great things.

**Fig. 4.3**   Lobelias growing on the slopes of Mount Kenya, Africa. The two tree-like plants are giant groundsels, *Senecio keniodendron*. The tall feathery spikes are *Lobelia telekii*, a semelparous species, about 2 m tall. The prickly plant in the right foreground with the shorter flower spike is another species, *Lobelia deckenii* ssp.*keniensis*. This is iteroparous and tends to dominate in moist valley bottoms (see distance) where adult mortality is low. Photo courtesy of Truman Young.

## 4.2   Invariants, combinations, and comparative studies

Researchers studying particular groups of organisms have long known that certain traits are correlated in very specific ways with other traits across species. The most famous of these correlations are those in the mammals, which display the so-called 'fast–slow continuum' (Promislow and Harvey 1990). Large-bodied mammals, such as elephants and buffalo, have long adult lifespans, take a long time to mature, have small litters of large offspring. In contrast, small-bodied mammals, such as mice and shrews, have short adult lifespans, mature very rapidly, and have large litters of small offspring. They therefore 'live fast and die young'. All mammals fit relatively neatly somewhere onto this continuum. We will examine how widespread this pattern is in other groups later on.

The differences just described are known because of comparative studies (Harvey and Pagel 1991), and much of the rest of this chapter relies on comparative studies. Comparative studies take variation among different species as their source of data. This variation has evolutionary origins, and much of the time the forces causing the variation we see among species have

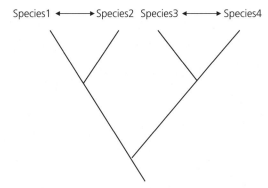

**Fig. 4.4**   Four species plotted onto an evolutionary tree, or phylogeny, showing their evolutionary relationships. The comparison between species 1 and species 2 is evolutionarily independent of the comparison between species 3 and 4. Both may be used as data in a comparative study.

been ecological ones. We can use the variation we see therefore as the results of some grand experiment in evolution played out over time, and ask what factors correlate with the variation. We have been using comparative data in a rather loose sense throughout the previous chapters: for example, we compared the species richness of different cichlid groups with different traits, the rates of recombination among species of different size, and the fecundities of annual verses perennial plants. More formally, comparative biologists like to use their data in some kind of statistical analysis, and the power of such analyses increases with the quantity of data. Normally, this means finding many different taxonomic groups that have changed independently of other groups. To ensure this is the case, comparative biologists often include in their study a 'phylogeny' that describes the relationships between the species in the dataset, and helps to identify independent evolutionary events (Figure 4.4).

Comparative studies on mammals have revealed the fast–slow continuum just described. The more specific details of this continuum are also interesting. First, many of the life history traits are related to body size across species by a characteristic **power function**, known as an allometric exponent (Figure 4.5). The birth rate of a mammal species, for example, is proportional to its body size to the power of $-0.25$ (thus, birth rate declines with body size across species). Amazingly, all of the allometric exponents for mammalian life history traits are close to some multiple of a quarter. This seems to require explanation. Furthermore, but rather less surprisingly given what we have just seen, certain traits, when multiplied together have values that are unrelated to body size. For example, if one trait with the exponent 0.25 is multiplied by another trait with the exponent $-0.25$, the result is a number that is not related to the body size of the species, and may be relatively constant. Eric Charnov, who has done most to bring these facts

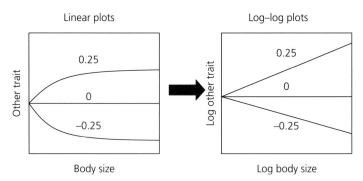

**Fig. 4.5** Allometry. Traits are allometric if they are related to body mass by a power function (linear plot, numbers indicate the exponent). If both the trait and body mass are logged, the relationship is a straight line where the slope of the line is equal to the exponent. Both 0.25 and −0.25 exponents are common in mammals. If a trait with an exponent of 0.25 is multiplied by a trait with an exponent of −0.25, the resulting product is an 'invariant' because it is unrelated to body mass (exponent of zero).

to the attention of other researchers, called these interesting numbers 'invariants'. In mammals, for example, one invariant is age at maturity times the adult annual mortality rate (this is in fact, relatively invariant within many groups of organisms but varies across groups). These strange facts seem to be telling us something pretty fundamental about life history evolution; the trouble is in knowing exactly what.

Charnov (1991) was the first to build a theoretical evolutionary model that would be able to predict these relationships. The basis of the model was natural selection on age at maturity governed primarily by adult mortality. One of the studies that most influenced this assumption is a study by Harvey and Zammuto (1985) using data on mortality rates of mammal species in their natural habitats. Getting data on that rather tricky variable for enough species has entailed centuries of man-hours in the field by dedicated researchers. Rather interestingly, the data showed that adult mortality rates in the wild strongly predicted where a mammal stood on the fast–slow continuum. Small-bodied mammals suffered high adult mortality and large-bodied mammals low mortality. Even more interesting, when body size was controlled for statistically, adult mortality rate was still related to most other life history traits: for example, mammals with low adult mortality for their size also had low fecundity for their size. This seemed to suggest that adult mortality might be a controlling selective force in shaping the fast–slow continuum.

In mammals offspring reach independence from their mothers at a certain size but then grow further until, when mature, they stop growing (Figure 4.6). Charnov then reasoned that the mortality rates they experienced after independence would govern when would be the best time to mature: if

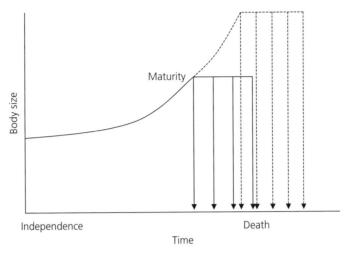

**Fig. 4.6**    Charnov's (1991) model. The solid line represents a mammal suffering heavy mortality after independence. It is best for it to mature early even though this means it can devote fewer resources to reproduction (vertical arrows). A mammal suffering lower mortality (dotted line) can afford to wait longer before maturing. As a result, it lives longer, is bigger when mature, and can devote more resources to reproduction. This model oversimplifies mammalian growth and metabolism (see Figure 4.8), yet makes accurate predictions across species.

they matured late, the benefit would be achieving a bigger body size, and consequent larger reproductive resources per unit time. However, this had to be traded-off against the risk of death before maturity. In general then, a late maturation time would only be selected if adult mortality was low.

Where do the allometric slopes in the model all come from? Charnov assumed a within-species allometry of metabolic rate on body size of 0.75. Before maturity, this metabolism causes growth, after maturity the metabolism is channelled into reproduction. Charnov had to make two more assumptions: first, about offspring size. He assumed that size at weaning is a constant proportion of maturation size across species. Finally, he assumed that the populations of mammals were of constant size, so that juvenile mortality balances fecundity. Density-dependent mortality acts only after the strategies have been decided on, so this does not contradict the idea of optimization. These assumptions are the origins of all the quarter-power scalings in the model.

The model can successfully predict the fast–slow continuum. For example, imagine an organism with low adult mortality. As a result of this, it lives a long adult life. It is selected to mature late and large because it can afford to delay reproduction to gain reproductive power due to the low risk of mortality before maturation (Figure 4.6). Since offspring size at weaning scales with body size to the power 1 in the model, but reproductive power only 0.75, it leaves fewer but larger offspring, which have a low risk of mortality prior to

**Table 4.1** Scaling exponents, relationships between variables and invariants predicted or assumed by Charnov (1991) and observed by Purvis and Harvey (1995)

| Variables | Theory | Observed |
|---|---|---|
| *Assumptions* | *Allometric exponent* | |
| Growth rate | 0.75 | 0.82✓ |
| Biomass offspring per year | 0.75 | 0.66✓ |
| Size at weaning to size at maturity | 1 | 0.89✗ |
| *Predictions* | *Allometric exponent* | |
| Age at maturity | 0.25 | 0.24✓ |
| Annual fecundity | −0.25 | −0.24✓ |
| Adult mortality rate | −0.25 | −0.24✓ |
| Juvenile mortality rate | −0.25 | −0.32✓ |
| *Pairs of traits* | *Relationship* | |
| Adult mortality, age at maturity | Negative | Negative✓ |
| Juvenile mortality, age at maturity | Negative | Negative✓ |
| Juvenile mortality, adult mortality | Positive | Positive✓ |
| Birth rate, adult mortality | Positive | Positive✓ |
| Birth rate, juvenile mortality | Positive | Positive✓ |
| Age at maturity times adult mortality | Size invariant | Size invariant✓ |
| Age at maturity times juvenile mortality | Size invariant | Size invariant✓ |
| Age at maturity times fecundity | Size invariant | Size invariant✓ |

*Note*: Ticks indicate that the data match the prediction, crosses indicate a mismatch.

independence. The model manages, via a bit of algebra, to predict nearly all of the allometric exponents and invariants seen in the data (see Table 4.1).

If we were to accept the model as an approximate description of the evolutionary forces at work, it would be a major achievement. It might, for example, explain the human life history (large body size, long adult lifespan, late maturation, few large offspring) in terms of low adult mortality in our evolutionary past. Other curious facts would be explicable. The longest lifespans of any mammal are achieved by the bowhead whale which typically lives to over one hundred years, and several may well have lived over 200 years (George *et al.* 1999). Such findings have raised some public interest: why do they live so long? If Charnov is right, then the answer is ridiculously simple: they live long because they experience low adult mortality. The model, of course, makes a number of simplifying assumptions, and several developments have since been made (Kozlowski and Weiner 1997; Charnov 2001, 2004).

How far do other organisms fit the Charnov model? One would think not many. For example, in many organisms, body size is not determined by a decision on when to mature, but by the size of a food patch allocated to them by their parents and the number of offspring with which they share it. Many insects, such as parasitic wasps and seed eating beetles, possess such a life history (Mayhew and Glaizot 2001). In others, such as altricial birds (which feed their offspring at the nest until fledging), offspring do not grow after independence at all but are raised to maturation size by their parents. In still others, such as most perennial plants, growth is not determinate, so organisms not only grow after maturation but also do not divert all their metabolic effort away from growth and into reproduction.

It would seem therefore that so many organisms break the assumptions of the model that it could not possibly be general. However, some of the model's predictions are turning out to be sufficiently general to suppose that some of the principles are common to many superficially different life histories. Flowering plants, for example, show many features of the fast–slow continuum predicted by Charnov's model, such as a negative relationship between adult mortality and age at maturity, despite breaking the assumption of deterministic growth (Franco and Silvertown 1997). Enquist *et al.* (1999) have even used Charnov's results for mammals to derive relationships between adult lifespan, age and size at maturity, and the density of wood across tropical forest tree species. These relationships are positive as predicted by the theory. Many indeterminate growers show some of Charnov's predicted invariants; parasitic nematodes for example, display an invariant maturation time multiplied by adult mortality rate, indicating that these are negatively related to each other with exponents of equal magnitude but opposite sign (Gemmil *et al.* 1999).

In contrast however, it is equally clear that it would not be valid to view all organisms as possessing mammal-like life histories. In birds, there is a general dichotomy between species that nest in safe places, such as albatrosses on offshore islands, and those that nest in dangerous places, such as grouse on the ground. The former suffer only low juvenile mortality, display low annual fecundity, grow slowly, breed late, and survive well as adults, while the latter display opposite traits. However, adult size is not consistently correlated with adult survival or with fecundity, but is robustly related to age at first breeding (Bennett and Owens 2002). These relationships are readily understood from the theory of single trait optimization. For example, high pre-breeding mortality, when maturation and growth are flexible, has long been known to select for high growth rates and early maturation, and early maturation selects for increased investment in reproduction and decreased investment in survival. In parasitoid wasps, there is no association between adult body size and either development time, adult lifespan, or fecundity (Blackburn 1991). Instead, these variables are correlated with traits specific to parasitoids: whether they develop as endo- or ectoparasitoids, and whether or not they permanently paralyze their host (Mayhew and Blackburn 1999). These traits probably exert their influence through constraints, such as on host range and egg size (Godfray 1994). Body size is related instead to host size and clutch size (see Mayhew and Hardy 1998). Thus, one of the exciting prospects of the next few years in life history evolution will be how many fundamentally different life histories there really are, and what the major selective influences are. Trying to understand invariant relationships, where they exist, is likely to be valuable.

## 4.3 The adaptive nature of metabolic scaling and its consequences

'I make no attempt to explain the 0.75 production scaling itself. I believe that it represents a fundamental coupling between an organism and its environment, but my attempts to derive it from even more basic considerations have thus far failed'. (Charnov 1993, p. 19)

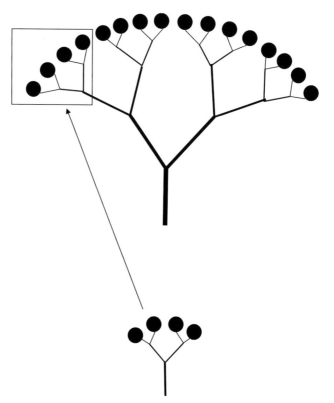

**Fig. 4.7** Optimal scaling of metabolic rate, following West *et al.* (1999). The problem is how to pack as many metabolic 'exchange units' of constant size (circles) into a three-dimensional volume (for simplicity the figure only shows two dimensions) while minimizing transport distance along the network. Two networks are shown: for a small organism (bottom) and a large organism (top). Note that the network for the small organism resembles a small piece of the network of the large organism. This 'self-similarity' or 'fractal' assumption helps generate the three-quarter power metabolic scaling. Metabolic rate is simply determined by the number of exchange units the organism contains, and is optimized with a three-quarter power relationship to body mass.

An unexplained assumption in Charnov's model was the three-quarter scaling exponent of metabolic rate with body mass. Recently, Jim Brown and co-workers have suggested an adaptive explanation for this (but see Kozlowski *et al.* 2003; Kozlowski and Konarzewski 2004), and have gone on to use it to not only explore many features of life history evolution but also other biological phenomena, including both ecological and evolutionary patterns, at large and small scales. The number of uses of the explanation is so large that Jim Brown and co-workers have to deny that it is a theory of 'everything', presumably settling for a 'large bundle of important things'. Work derived from this original theory even has its own name, dubbed 'metabolic ecology'.

The three-quarter scaling is most easily understood as arising from a packaging problem (West *et al.* 1999). Imagine that the total area provided by exchange units of constant size limits metabolic rate. These units might be alveoli in the lungs, capillaries, leaves, or root hairs. We want to know how the density of these units should scale with body mass so as to maximize metabolic rate while minimizing the delivery distance from the surface to the point of demand and while being packed into a three-dimensional volume (Figure 4.7). According to West *et al.*, this occurs when the number of exchange units is proportional to mass to the three quarters. The maths, for those so inclined, is explained in Box 4.1.

---

**Box 4.1**  The surface area of an object is proportional to its mass to the power 2/3, because area is two-dimensional and mass is three-dimensional. The area of exchange surfaces *within* a volume scales with slightly greater freedom provided by two parameters, one relating to the number of exchange units ($\varepsilon_a$) and one to the distance between them ($\varepsilon_l$):

$$a \propto M^{(2+\varepsilon_a)/(3+\varepsilon_a+\varepsilon_l)}.$$

In a three-dimensional shape like a body, these parameters can vary between 0 and 1. It is easily seen that $a$ is maximized when $\varepsilon_l = 0$ and $\varepsilon_a = 1$. This makes sense because when $\varepsilon_a = 1$ the area is maximized and when $\varepsilon_l = 0$ the distance between exchange units is minimized. When these values are taken, the above becomes

$$a \propto M^{3/4}.$$

If metabolism depends directly on area of exchange units, it takes on this three quarters exponent.

**Table 4.2** A few theoretical predictions that have arisen out of the three-quarter scaling of metabolic rate and body mass

| Phenomenon | Assumptions | Predictions | Reference |
|---|---|---|---|
| Scaling of plant population density | Plants compete for resources and grow until limited by them; resource supply per unit area is constant | Scaling of plant density with size is −4/3; energy use per area is therefore size invariant | Enquist et al. (1998) |
| Scaling of production and life history traits in plants | For a given mass, stem diameter depends on wood density; proportional investment in reproduction is constant; maturity and lifespan determined by assuming determinate growth | For trees of the same diameter, production is independent of wood density, relative growth rate scales as −0.25, trees with dense wood live longer and mature later | Enquist et al. (1999) |
| Above ground structure of plants | Plant design is limited by branching supply networks; trunks and branches resist buckling while minimizing energy dissipation | Scaling of leaf number (0.75), height of tree (0.25), flow rate per tube (0). Maximum tree height is about 100 m, limited by the required width of the supply tubes | West et al. (1999) |
| Ontogenetic growth rates | Growth depends on metabolic supply and also on the demands of maintenance | Growth is sigmoidal, and at a given proportion of asymptotic mass, all organisms spend about the same proportion of metabolism on growth | West et al. (2001) |
| Latitudinal gradient in species richness of guilds across communities | Total energy flux per area is body size invariant, number of individuals in guild is constant | Number of species increases with temperature in ectotherms | Allen et al. (2002) |
| Metabolic rates in mammals, cells, mitochondria, and molecules | Cells within a body, mitochondria within a cell, and respiratory complexes within mitochondria are all supplied by fractal-like space-filling networks | Knowing metabolic rate at one level of organization is sufficient to predict the others | West et al. (2002) |

Thus natural selection may provide an explanation for the scaling, and hence, yet another contribution to the explanation of mammalian life histories, as well as other quarter-power scaling exponents that derive from it. However, combined with other assumptions, it can be used as the starting point for the explanation of many other phenomena, most of which are supported by comparative data. Table 4.2 lists some of these. I cannot help but quote Churchill again: 'All the great things are simple, and many can be expressed in a single word'. In this case, the word is 'quarter'.

Let us see how one of these arguments works for a life history trait: growth rate (West *et al.* 2001). The aim is to predict how animals grow from basic assumptions about how metabolic rate scales with body mass. First, assume that growth is fuelled by metabolism, and occurs by cell division. Then assume that during growth a proportion of metabolism is devoted to making new tissue and a certain amount to maintenance. The addition of new mass depends on supply, proportional to mass to the three quarters (how the units of exchange scale), and on demand, which is directly proportional to mass (the number of cells). Thus, as growth occurs, and mass increases, demand begins to outstrip supply, ultimately limiting growth (Figure 4.8).

This is beguilingly simple. If this simple model of growth is correct, it can be shown that all animals should fit a single sigmoidal curve equation, and this equation contains no less than three quarter-power exponents resulting from the original metabolic assumptions. The values of some of the equations' parameters vary from species-to-species and this is what gives different organisms apparently different growth profiles. Data for several vertebrates do indeed fit the predicted curves, although in practice they cover different parts of it: at birth a cow is already expending half its energy in maintenance, and quickly approaches asymptotic size. A cod, however, only spends half its energy on maintenance after 6 years of age. The cod can

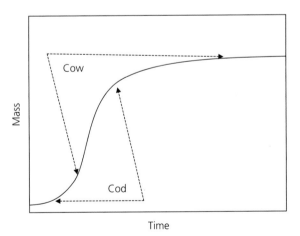

Time

**Fig. 4.8**   The sigmoid curve of animal growth and development, following West *et al.* (2001). The curve is asymptotic simply because demand increasingly outstrips supply as size increases. The underlying equation of the line is the same for all species. However, because some of the parameters take species-specific values, the curves would appear to differ for different species when plotted on the same axes shown. Cod live most of their lives on the left part of their curve maturing well before the asymptote (indeterminate growth), and cows mostly on the right, maturing closer to it (determinate growth). Compare this with Figure 4.6.

never really expect to reach asymptotic size, and in turn devotes a substantial proportion of total lifetime energy to growth (about 40%). A cow, however, expends only about 1% of its lifetime energy use to growth, and only 10% of its pre-maturation energy use. Thus, according to the model determinate and indeterminate growth simply reflect whether an organism reaches asymptotic size before death.

The model, even more so than Charnov's model, emphasizes the similarities among organisms rather than their differences, something that applies to most of the models that rely on common scaling assumptions. The model is mechanistic rather than evolutionary, and merely invokes natural selection for the original metabolic scaling assumption. Rather nicely, Charnov (2001), following on from the above work, has incorporated these new growth assumptions into a modified theory of mammalian life history evolution. The new model is able to retain the original quarter-power scaling predictions of his original model (Charnov 1991) but now makes more realistic assumptions about growth and metabolism that themselves derive from first principles. Life history theory, therefore, traditionally concentrating on individual traits, moving into invariant relationships and a more whole organism view, has come full circle again. Throughout however, the big message the field brings is that natural selection has played a fundamental role in generating both the basic similarities that organisms share, and the differences that make each unique. The field has made exciting claims in recent years about how evolution shapes us all, and there is more excitement to come. The following chapters consider traits that do not find traditional placement in life history evolution, but might easily have done so.

## 4.4   Further reading

Optimization theory and comparative studies are introduced by Krebs and Davies (1993). Stearns (1992) and Roff (1992) summarize traditional life history theory, and Lessells (1991) provides a useful quicker overview. The mammal story is introduced by Harvey and Purvis (1999), after which try Charnov (1993) for a broader overview to invariants. Brown and West (2000) summarize some of the scaling work, and special issues of *Functional Ecology* **18**(2) and *Ecology* **85**(7) debate some matters arising.

# 5 Sons, daughters, and distorters

Most of us take for granted three facts about our gender: (1) that we each produce only male or female gametes, not both; (2) that about half of us are male and half are female; and (3) that this is a result of the mechanism of sex determination known as male heterogamety. In male heterogamety, possession of two similar (X) chromosomes determines a female individual, whereas possession of two non-identical chromosomes (X and Y) determines a male. Because haploid sperm are produced by fair (Mendelian) segregation of the male sex chromosomes, half produce daughters and half sons. This keeps the sex ratio at approximate equality.

In the natural world as a whole however, none of these facts may be taken for granted; many organisms are **cosexual**, or display biased sex ratios, or have different sex determining mechanisms. Instead of assuming the above three facts, we should instead ask the following three questions:

1. What makes an organism **dioecious** as opposed to cosexual?
2. What favours equal or biased reproductive investment in the sexes?
3. How do sex determining systems evolve and why?

This chapter is primarily about answering these three questions. Sex allocation (the first two questions) is often considered the 'jewel in the crown of evolutionary ecology' (West and Herre 2002). It was one of the first fields to clearly match a knowledge of ecological circumstances to evolutionary outcomes. For this reason alone it must be discussed. However, recent advances are suggesting a far more interesting role for sex allocation: as a driver of other evolutionary and ecological phenomena. The chapter will explore this possibility as well.

## 5.1  Male and female in one individual or separate individuals?

Let us start with the first question. We will take a broad fitness-based approach first outlined by Charnov *et al.* (1976). They viewed the question of whether to express both sexes in the same individual or in separate individuals as analogous to asking if an organism should specialize in one role or

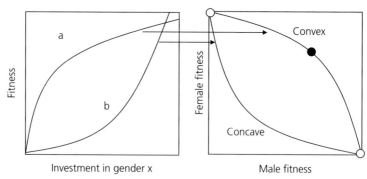

**Fig. 5.1**  The evolution of hermaphrodites, following Charnov *et al.* (1976). On the left are fitness curves detailing how much fitness derives from a given investment in resources. Curve 'a' is asymptotic, meaning that each extra investment gives less extra fitness return, whereas curve 'b' accelerates, meaning that only a high level of investment in that sex will return much fitness. On the right are fitness trade-offs between the sexes. Curve 'a' translates into a convex trade-off, meaning that switching investment to the alternative sex does not greatly reduce the fitness of the other sex. For such curves, hermaphroditism is optimal (black circle), because it gives the greatest total fitness derived from both sexes. If the fitness curve accelerates (b), trade-offs become concave, and the optimum is to be either male or female but not both (open circles).

be a generalist, assuming more than one role. We will encounter this problem again in a more ecological context in Chapter 9, so the discussion here serves as a useful introduction. First, imagine a trade-off between investment in male versus female fitness. This can be a linear trade-off, convex or concave (Figure 5.1). It is simple to show that when the curve is convex, cosexuality is stable: a pure female, reallocating resource to 'male-ness', gains more fitness than it loses, because the cosexual is almost as good a female as the pure female itself while there is some male-ness added that also gives a fitness return. Similarly, a pure male allocating resources to female-ness would be almost as good as being male, and yet, get a bit of fitness through female function too. What might cause the trade-off to be convex? We can understand this from an examination of the relationship between investment in one of the sexes and fitness. If the added fitness from extra investment in one sex tends to decline, selection will increasingly favour investing in the other sex (Figure 5.1). This means that just a bit of investment in either sex will make the plant both a perfectly good male and a perfectly good female, and the fitness trade-off will be convex.

The distribution of cosexuals in plants makes sense in this framework. Cosexuality, for example, is associated with animal as opposed to wind pollination, and with wind as opposed to animal-dispersed seeds (e.g. Bawa 1980; Charnov 1982; Vamosi *et al.* 2003). All these associations can be related to the shape of the fitness gain curves. For example, wind-dispersed seeds need to be light and therefore fitness asymptotes strongly with extra investment. Animal-dispersed seeds, however, need to be resource-rich to attract animals, and may even display an accelerating gain curve (Figure 5.1). Thus, selection favours dioecy in animal-dispersed plants, and cosexuality in wind-dispersed plants. With pollination, however, it is the reverse: successful wind pollination demands pollen en mass, thus, the male gain curve is unlikely to saturate at all strongly, thus selecting for dioecy. Although this approach can claim some intuitive success, theory of the evolution of cosexuality verses dioecy remains poorly tested in general: it is very challenging to measure the gain curves necessary to confirm or refute hypotheses, and many hypotheses are consistent with a single set of gain curves (see Charnov 1982). We will encounter similar problems in later chapters.

## 5.2 Sex allocation bias

A second sex allocation problem is understanding bias in allocation towards one of the sexes overall, such that in dioecious populations there is a biased sex ratio, or in cosexuals greater investment in one sex function. Our understanding of this problem originates largely with Fisher (1930) (though see Edwards (1998); Seger and Stubblefield (2002)). Fisher pondered the reasons why organisms tend to display 50 : 50 sex ratios. His argument is one of frequency-dependent mating success. Consider a gene that biases offspring production towards one sex. In the next generation that sex will find itself more common, and will on average have lower mating success than the opposite sex, hence, production of grand-offspring is reduced. Selection will favour genes that bias the sex ratio in the opposite direction, until the sex ratio reaches 50 : 50 again. An equal sex ratio is therefore an Evolutionarily Stable Strategy (ESS) because selection will immediately tend to counter any bias. This verbal argument of evolutionary stability is slightly different to the fitness optimization problems mentioned in the last chapter in that the fitness of any strategy depends on the strategies adopted by the rest of the population, so the optimum shifts as evolution proceeds. This presents novel challenges when attempting to predict evolutionary outcomes, and these have led to a range of theoretical innovations.

Observing Fisher's equalizing selective process in action has not been widely possible because of a general absence of genetic variation in the sex ratio in organisms displaying a 50 : 50 sex ratio. However, a small marine fish, the Atlantic Silverside (Figure 5.2), showed its practical validity for the first time (Conover and Vanvorhees 1990). In this fish, the temperature at which offspring develop determines sex. Since the fish has a large geographic distribution, the temperature that results in an equal sex ratio is

**Fig. 5.2** The Atlantic silverside, *Menidia menidia*, which displays environmental sex determination under frequency dependent selection, thus maintaining a equal sex ratio. Photo by R. George Rowland and provided courtesy of David Conover.

lower in higher latitudes where the water is colder. In an elegant experiment, populations of fish from different regions were reared at abnormal water temperatures, and the population sex ratio was naturally biased, but over time the sex ratio equalized because of selection on the temperature threshold at which sex switches. This is exactly the process Fisher envisaged, showing it works in nature as well as on paper.

In the general absence of studies like this, the validity of Fisher's argument has been largely tested indirectly from the association between an observed sex ratio bias and the breaking of one of Fisher's implicit theoretical assumptions. Charnov (1993) has extended his ideas of 'invariance' relationships (Chapter 4) by reference to Fisher's assumptions: the 50 : 50 sex ratio can be understood as a life history invariant resulting from two underlying 'symmetries'. The first is that of inheritance: an offspring receives half of its genes from its father and half from its mother. This tends to make both male and female offspring equally valuable because half of any offspring's genes will come from males and the other half from females. If one sex ever became a more profitable way of transmitting genes than the other, this symmetry would be broken and there would be the potential for sex ratio bias. The second symmetry is that of proportional gains through sons and daughters: when mothers switch investment from one sex to another, their reproductive gains and losses are constant for all mothers and in all circumstances. The alternative is that certain mothers may gain more fitness than others from investing in a particular sex.

## 5.3   Inheritance asymmetry

The inheritance symmetry is known to be broken in two contexts. First, some **nuclear** genes, such as those on the sex chromosomes, are transmitted more effectively through one sex (Hamilton 1967). Y chromosomes that are able to bias the sex ratio towards the heterogametic sex (XY males in mammals) should be favoured over those that do not, because the homogametic sex (XX females in mammals) never transmits them. Similarly, the X chromosome will be favoured by a bias towards the homogametic sex, but not as strongly since it is also transmitted through the heterogametic sex (only half as efficiently).

These so-called driving sex chromosomes, or meiotic drivers, are now known from a number of organisms. Most, like 'segregation distorter' in *Drosophila* fruit flies, are X chromosomes that cause female bias. One possible reason is that Y chromosome drive is likely to be very strong and may cause extinction of a population (through elimination of one sex) before counter-measures can evolve. Driving Y chromosomes will therefore usually not persist long enough to be observed. Countermeasures may come from modifier genes on other chromosomes that suppress the action of the driver gene. In the fly *Drosophila simulans*, X-linked drive is suppressed by modifier genes on the Y chromosome and all the major autosomes (see Capillon and Atlan 1999).

Other, non-nuclear genetic elements can also select for sex ratio bias. Mitochondria are cytoplasmic organelles originating from symbiotic bacteria (see Chapter 2). They are inherited only through eggs which carry the cytoplasm for the zygote. In plants, mitochondrial mutants are responsible for the phenomenon known as cytoplasmic male sterility (CMS), whereby otherwise cosexual plants become entirely female (see Saumitou-Laprade *et al.* 1994). CMS is common in agricultural plants, such as maize, rice, sunflowers, and beans. Many insects similarly possess a symbiotic bacterium called *Wolbachia*, which is cytoplasmically inherited (Figure 5.3). *Wolbachia* can have several effects on insects, including the production of all-female lines (parthenogenesis). Incredible evidence for this first came when males were discovered for the first time in some parasitic wasp species after adding antibiotics to their diet, which killed the *Wolbachia* (Stouthamer *et al.* 1990). Another class of maternally inherited microbes are known as male-killers because of the way they achieve female biased sex ratios. They have been found in species in a number of insect orders, including wasps, beetles, butterflies, and flies. *Wolbachia* can act as a male killer in some insects and is responsible for producing the most extreme sex ratio biases known: in the butterfly *Hypolimnas bolina*, on Upolu island in Samoa (Figure 5.3). Ninety-nine per cent of individuals are female in this population (Dyson and Hurst 2004).

(a)

(b)

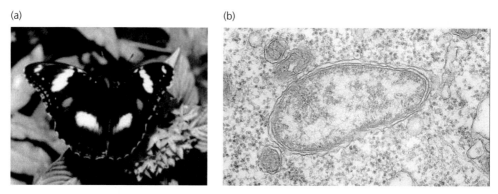

**Fig. 5.3** (a) The butterfly *H. bolina* (wingspan 7 cm), whose sex ratios are extremely female biased due to male-killing. (b) *Wolbachia* (length 2 $\mu$m), shown here in an *H. bolina* egg. Photos courtesy of Greg Hurst.

Therefore, breaking the one-father one-mother symmetry frequently leads to sex ratio bias, as well as conflict between different genetic elements over the sex ratio. This is important as will be seen later in this chapter and later in the book, because conflict of interest may be a powerful driver of evolutionary change.

## 5.4   Proportional gains asymmetry

The second Fisherian symmetry is that of proportional gains from sons and daughters. There are two famous ways in which asymmetric gains can result in biased sex allocation. The first was realized by Hamilton (1967). Many organisms display a 'subdivided' mating structure, meaning that females obtain their mates from a relatively local pool of males. In such groups there is competition among males for mates, and mothers contributing offspring to the mating pool will achieve greater production of grand-offspring if they bias their production of offspring towards females. The reasons are two-fold: first, they do not waste effort on sons with low mating success; second, they instead produce females that will be mated (Figures 5.4, 5.5). In populations with less spatially structured mating (i.e. more females contribute offspring to each mating pool), sex allocation should approach equality. This is observed in numerous wasp species that lay their eggs on resource patches where the offspring also mate (hosts insects in the case of parasitoids, figs in the case of fig wasps) (see Werren 1983; Herre 1985). In plants that habitually self-pollinate, there should be similar selection pressure for a reduction in male allocation. There is comparative evidence from numerous plants that allocation to male function decreases with the selfing rate (Charnov 1982).

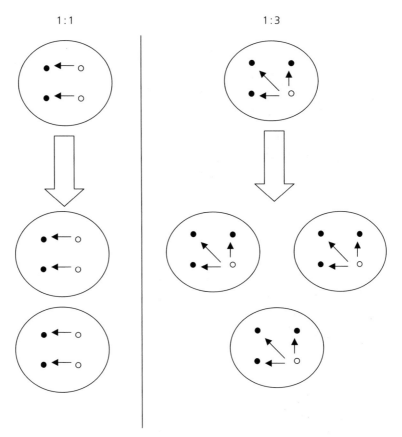

**Fig. 5.4** The evolution of female biased sex ratios under local mating. Two situations are shown, one with a 1 : 1 male to female ratio (left), and one with a 1 : 3 ratio of males to females. Large circles represent offspring groups which mate among themselves before dispersal to other patches. Closed small circles are females and open small circles are males. Small arrows represent mating. Under a 1 : 1 sex ratio, two mated females disperse to form two patches of offspring in the next generation. Under a 1 : 3 sex ratio, however, three females disperse, leaving more descendents. Thus, selection favours a female bias.

The second significant way in which asymmetric gains can emerge is when one sex benefits more from investment than the other sex in some environments (Trivers and Willard 1973; Charnov *et al.* 1981) (Figure 5.6). Many polygynous cosexual fish, for example, are female early in life, and switch to become males later in life when they are also larger. Presumably, taking over and defending a harem of females is only possible for males if they are large, but it is then highly profitable. Females, however, can produce offspring even when small. Many cosexual plants also bias their allocation towards pollen and away from ovules under conditions of stress, presumably because pollen is small and much less resource-dependent

**Fig. 5.5** Subdivided mating structures in nature. Pollinating fig wasps, *Pleistodontes froggatti*, inside a Moreton Bay fig, *Ficus macrophylla*, that has been cut open. The wasp in the right foreground is laying an egg into a fig flower. The female wasps have no wings as these are lost while pushing their way into the fig through the narrow entrance. The offspring of these females will mate with each other inside the fig; hence there is local mate competition and selection for a female biased sex ratio. The full width of the photo is approximately 2 cm. Photo courtesy of James Cook.

than fruit. Of course, underlying all these sex allocation biases is a sex determining mechanism. Next we will investigate how these mechanisms might evolve.

## 5.5    The evolution of sex determining systems

Sex determination systems are highly variable across taxa. Heterogamety has already been mentioned. Actually, this is a general term that conceals much underlying genetic diversity. Males are the heterogametic sex in mammals, females in birds, but reptiles, amphibians, and fish display both variants, often within the same family. Different genetic systems are even known to underlie male heterogamety: in mammals dominant male determiners exist on the Y chromosome. In common sorrel plants, the Y chromosome is inert, and gender is determined by the ratio of X chromosomes to **autosomes**. In the Hymenoptera and some other arthropods, haplodiploidy occurs, whereby males are haploid, developing either from unfertilized eggs or from

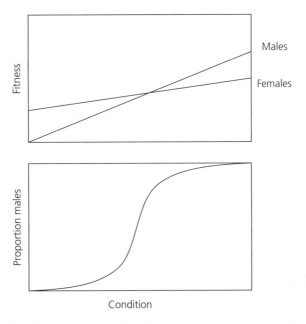

**Fig. 5.6**   The Trivers-Willard hypothesis of conditional sex expression. In organisms with polygynous mating systems (mammals, some fish), males benefit more from good condition than females (top). Therefore selection favours male production under good conditions and female production under poor conditions (bottom). In other organisms (most plants, many insects), females benefit more from good conditions and sex allocation is reversed.

loss of the paternal genome after fertilization. Females are 'normal' diploids. In addition to such 'genetic' systems are the so-called 'environmental' systems whereby key variables experienced during development determine gender. For example, in many reptiles it is temperature that determines whether an egg becomes male or female. However, in some lizards and alligators, males develop at high temperatures, while in some turtles it is the other way around. In other turtles and crocodiles, males develop only at intermediate temperatures. In the European pond turtle, both genetic and environmental factors determine sex. How did this tremendous variety evolve? This question is one in which researchers are still just beginning to make headway, but the results so far are fascinating.

In principle, since equal sex ratios are often expected to be evolutionarily stable, so will mechanisms that tend to result in equal sex ratios. This phenomenon, whereby selection favours genes that lead to particular sex ratios, is known as sex ratio selection (Chapter 1), and is also the mechanism by which sex allocation evolves. Heterogamety, because of **Mendelian inheritance**, will tend to result in equal sex ratios, hence, should often be selected for. Heterogamety appears to be relatively conserved (it is the exclusive mechanism in both birds and mammals, both groups with a long ancestry).

Because sex determination and sex allocation both evolve through the same broad mechanism of sex ratio selection, they might be expected to evolve in response to the same environmental factors. For example, it is widely supposed that haplodiploidy has been selected for in insects and mites by the presence of a subdivided mating structure (i.e. a proportional gains asymmetry), the same force that selects for variable sex ratios in subdivided populations, by allowing mothers ease of behavioural control over sex allocation. Environmental Sex Determination should also be favoured when offspring quality is differentially affected in the two sexes by factors that act during development, the same force that selects for conditional sex expression (Charnov and Bull 1977).

Inheritance asymmetry may also select for changes in the sex-determining mechanism through conflict between genetic elements over the sex ratio. We have already seen that different genetic elements may differ over the preferred sex allocation strategy. For example, sex chromosomes should favour biased sex ratios while autosomes should generally favour equal sex ratios. Similarly, cytoplasmic elements, such as mitochondria, should favour female biased sex ratios, conflicting again with autosomes. What effects do these conflicts have? One immediate effect is that if a force other than the parental autosomes takes control, this can result in a novel mechanism of sex determination. Under cytoplasmic male sterility in plants, sex is determined not by the ancestral mechanism, but by the presence of mitochondrial mutants and restorer genes which determine whether an individual is cosexual or female. In populations of *Plantago lanceolata* (Figure 5.7) there are three known CMS mutants and three restorer genes known (de Haan *et al.* 1997). Some populations have become fixed for one particular mutant and restorer, to the extent that hybrids between populations are sterile. This is rather interesting as it suggests that conflict over sex allocation can lead to reproductive isolation, and potentially even to speciation. The reason conflict can have these effects is that there is selection on both parties in the conflict to rapidly evolve countermeasures such that they have the upper hand in the conflict (an arms race). This can lead to rapid genetic differentiation between populations evolving in isolation.

The other interesting effect of conflict between autosomes and mitochrondria in CMS systems is that it may facilitate the transition from cosexuality to dioecy. Taxonomically, there is an association between the presence of dioecy and gynodioecy (individuals either cosexual or female). Furthermore, some species appear to have evolved towards dioecy as a result of CMS. When some individuals in a population are fully female as a result of CMS, there can be selection of cosexuals to reduce their allocation to female function in order to restore the population equilibrium of equal allocation. In the common

(a)

(b)

**Fig. 5.7** Cytoplasmic male sterility in *P. lanceolata*. (a) A normal hermaphrodite flower with both stigmas and **anthers**, and (b) a male-sterile lacking anthers. Photos courtesy of Hans Peter Koelewijn.

thyme, *Thymus vulgaris*, cosexual individuals in populations with CMS, indeed, bias their allocation towards males in this way (Atlan 1992). Thus, conflicts as a result of inheritance asymmetry can not only cause biased sex allocation, but perhaps also shifts from cosexuality to separate sexes.

Another interesting example of how conflict over sex allocation can cause the evolution of sex-determining systems comes from the common wood-louse, *Armadillidium vulgare* (Figure 5.8). This has an ancestral system of female heterogamety, termed ZW females, ZZ males. Some females, however, are infected with *Wolbachia* and produce an excess of females. The *Wolbachia* actually works by converting ZZ males into females (ZZ + *Wo*) (Rigaud 1997). Several field populations now lack the normal W chromosome entirely and consist of ZZ males and ZZ + *Wo* females. At this stage, these populations have evolved from a normal chromosomal sex-determining system to a cytoplasmic one. A further feminizing factor was then discovered, labelled *f*, in a population derived from a single ZW female formerly inoculated with *Wolbachia*. However, it emerged after failure of *Wolbachia* transmission, suggesting that it might be an autosomal countermeasure to *Wolbachia*. *f* can also sometimes become fixed on a male chromosome (Z). This effectively then becomes a new female (W*)

**Fig. 5.8**    Male (right) and female (left) *A. vulgare*. Photo courtesy of Didier Bouchon.

chromosome. Thus, conflict can lead to restoration of the chromosomal system again (Figure 5.9). The woodlouse system shows how conflict can lead to potentially rapid turnover in sex determination.

## 5.6   The jewel in the crown?

We began this chapter with three questions. In brief, how have we answered them? First, separate sexes may evolve when an individual functions better as a single gender than as two. Second, biased sex allocation may evolve when one of two symmetries are broken. Finally, sex-determining systems probably evolve under the same processes and circumstances as sex allocation. We also began this chapter with the statement that sex allocation is the jewel in the crown of evolutionary ecology. We have seen that in one respect it is not, for our understanding of the evolution of cosexuality and dioecy, while certainly promising, is not as advanced empirically as our understanding of sex allocation bias. Both sex allocation and sex determination probably evolve by a common mechanism, and under common ecological circumstances. Sex allocation takes place in the context of sex determination, and sex determination can evolve in response to sex allocation decisions. Our

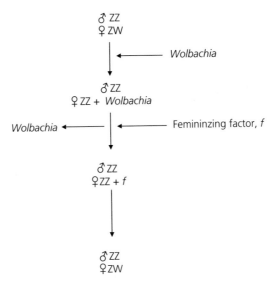

**Fig. 5.9**    Evolution of sex determination in *A. vulgare* (after Rigaud 1997).

understanding of the evolution of sex determination is not far advanced, but an integrated study of sex allocation and sex determination may yet prove to justify a central place in evolutionary ecology, not simply because it is a shining example of what is possible, but because of numerous potential impacts on ecology and evolution, such as population extinction and in reproductive isolation. Sex allocation, from Hamilton's extraordinary example, illustrates the power of genetic conflict to change living systems.

## 5.7    Further reading

Excellent short summaries of sex allocation are by Frank (2002), Policansky (1987), and other articles in the same issue, and Charnov (1993, ch. 2). Hardy (2002) provides a useful series of reviews on many sex allocation and determination issues. For an approachable text about sex ratio distorters try Majerus (2003). Werren and Beukeboom (1998) review sexual conflict in sex allocation and sex determination. Charnov (1982) and Bull (1983) are classic works, full of good stuff, but require confidence with the broad issues.

# 6 Voyagers, residents, and sleepers

There are two lasting bequests we can give our children. One is roots.
The other is wings.

Hodding Carter Jr

One of the most important developments in evolutionary biology came in
1962 with the publication of 'Animal dispersion in relation to social behaviour' by Vero Wynne-Edwards. This was probably the most controversial
biology book of its generation. In it, Wynne-Edwards proposed that animal
populations were regulated by dispersal of surplus, less fit, individuals,
which prevented overpopulation. Wynne-Edwards believed this behaviour
to have evolved through differential survival and extinction of populations;
those groups of individuals without this benevolent behaviour went extinct,
leaving the remaining habitat to become populated by those that had
(Figure 6.1). This process, group selection, ran counter to the Darwinian
notion of selection between individuals, and was robustly rejected by
the academic community. The principal legacy of the book was to focus
attention on the mechanisms of natural selection and the evolution of
social behaviour. The ideas that followed laid the foundations of modern
evolutionary ecology.

   Though Wynne-Edwards's ideas on group selection rightly never gained
acceptance, the questions on which his work was based remain as some of the
most challenging in evolutionary biology. Many animal and plant species
display risky and costly dispersal behaviour. Having been born in a habitat
that was obviously suitable for their parents, they undertake movements to
new breeding habitats, exposing themselves to predators and unsuitable
environments, often at great energetic cost, and with no guarantee of success.
How could such behaviour evolve? In particular, can we explain it in the standard Darwinian context of selection on individuals, rather than invoking
group selection? The issue is, as Wynne-Edwards appreciated, a very important one. Dispersal is the process that binds the populations of a species
together. It can not only regulate the dynamics of populations, but also their
genetic differentiation, and coexistence and evolution with other species

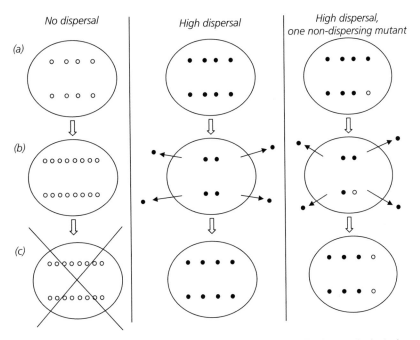

**Fig. 6.1**  Wynne-Edwards's view of the evolution of dispersal by group selection, and why it does not work. In the left column is a population with no dispersers (a), which becomes overpopulated (b) and goes extinct (c). A population with high dispersal propensity (middle column) would thus survive, even though the dispersers have very low survival chances. This was Wynne-Edwards's argument. However (right column), such populations are susceptible to cheats (open circle), with a low dispersal propensity. Cheats do not pay the cost of dispersal but gain the benefits from the rest of the population, and hence increase in frequency.

(see Chapter 11). Understanding why it evolves might unlock our understanding of a whole gamut of evolutionary and ecological phenomena.

In addition to dispersal via their seeds, plants display an equally puzzling but related phenomenon: seed dormancy (delayed germination). There are two obvious costs to dormancy. First, there is some year-to-year mortality of dormant seeds, so the total number of germinating seeds is reduced. Second, in increasing populations there is selection for early reproduction, giving more descendents by virtue of more generation cycles. Dormancy delays reproduction. Hence, it is another costly and curious phenomenon. And seed dormancy, like dispersal, is a population parameter that can affect dynamics and coexistence. Rather nicely, but with hindsight not surprisingly, the evolutionary solutions to dormancy and dispersal are intricately related at a theoretical level, and, in plants at least, can no longer be considered in isolation.

# 6.1   Evolutionary stable dispersal strategies

One of the great developments in evolutionary biology that resulted from Wynne-Edwards's challenge was the notion of the evolutionary stable strategy or ESS (Chapter 5). These are strategies that, when adopted by a population, cannot be invaded by any alternative strategy. ESSs result from the relative fitness of individuals displaying alternative strategies. They are used to predict evolutionary outcomes when the fitness of a given strategy depends on the strategies adopted by the remainder of the population.

Hamilton (1967), in his paper on extraordinary sex ratios, is normally attributed with the first explicit ESS model in evolutionary biology. It was Hamilton again who, together with Bob May, first applied the approach to the evolution of dispersal (Hamilton and May 1977). Hamilton and May imagined a very simple scenario in which the environment consisted of a number of identical habitat patches that were each occupied by a single individual. These adults died every year (an annual species), and the offspring then competed to exploit the vacant patches. Adults would give rise parthenogenetically to dispersive offspring, with a given frequency. Dispersing offspring were then allocated evenly across the patches, while non-dispersing offspring remained in the natal patch to compete. The outcome of competition within a patch was random, and dispersive offspring suffered a mortality cost before arriving at their new patch.

Given these conditions, what is the ESS rate of dispersive offspring? The solution turns out to be remarkably simple: $1/(2 - s)$ where $s$ is the survival of dispersers from their natal patch to the new patch. Thus, if there is no mortality cost to dispersal, all individuals disperse. If the mortality cost is extremely high ($s \rightarrow 0$) over 50% of individuals should still disperse! Why should this happen? Imagine a scenario in which the resident strategy is for zero dispersal (Figure 6.2). A mutant individual with some level of dispersal will displace this strategy: its offspring will compete for, and some of the time gain, new patches which the resident non-dispersers can never regain. The specific genetic advantage from dispersive offspring is that offspring compete less among themselves and more among alternative strategies which they can displace; there by reducing competition between kin that share the same genes.

When they added in the possibility that patches could become totally vacant (by distributing offspring randomly, not evenly among patches), Hamilton and May noticed an interesting point: the ESS dispersal strategy was always greater than that which would maximize site occupancy. The implications are, first that the predominance of individual over group selection increases dispersal, and second that what is best for the individual is not always best for the population. This second theme is actually one of the most

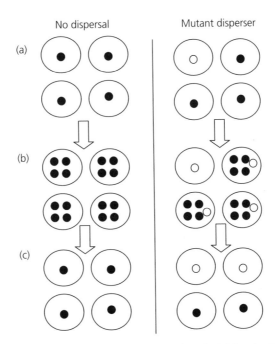

**Fig. 6.2** Hamilton and May's model of the evolution of dispersal. In the left hand column, individuals (small circles) are distributed among four patches (large circles). After reproduction (b), offspring compete and mortality acts (c). In the right hand column, a mutant disperser (small open circle) arises in one patch (a). Some of its offspring disperse to neighbouring patches (b), and some of them are successful. After competition, its patch occupancy has increased, hence, dispersal tendency spreads through the population.

important messages of the evolution of dispersal, one that has been reinforced by subsequent work. In fact dispersal evolution can theoretically both increase the risk of population extinction and reduce it (see Chapter 13).

Since then, a number of authors have extended Hamilton and May's assumptions, normally recovering their basic result as a special case (see Johnson and Gaines 1990). It was clear, however, that the mathematical complexities of the ESS approach would sometimes become insurmountable, and many subsequent authors have taken alternative modelling approaches, such as evolutionary simulations (see Chapter 1). Comfortingly these models have tended to reinforce the findings of the ESS models, such that dispersal theorists can present to the world an almost united front. What have they found?

First, variation in site suitability over time tends to favour dispersal. This suggestion preceded the advent of formal modelling (e.g. Southwood 1962) but has been confirmed by it. Temporal variation rewards individuals that hedge their bets by placing offspring in a variety of patches in case the current patch deteriorates. A special case of this is the so-called 'Janzen–Connell' hypothesis (Janzen 1970; Connell 1971), usually used to explain species coexistence. In some organisms local habitat patches may become unsuitable simply because it is already occupied by the species. One reason may be because parents harbour natural enemies that can affect offspring, and the effect of those enemies decreases with distance from the parent. This selects strongly for dispersal between individuals of the same species, and hence, also generates a species-rich local mix of individuals. Chaotic local population dynamics, which also make patch quality unpredictable in time, can select for dispersal in the same way.

Spatial, rather than temporal, variation in the environment, has the opposite effect on dispersal evolution: it will tend to reduce it because most individuals will be in the best sites, so fewer individuals benefit from moving from poor to good sites than lose from moving from good to poor sites. This spatial variation includes habitat fragmentation, which can reduce dispersal rates severely (Travis and Dytham 1999). Clearly, in a situation where populations persist through migration between subpopulations (a metapopulation), habitat fragmentation will select for reduced dispersal. This will enhance the probability of extinction, a nasty side effect of selection acting on individuals rather than groups. This phenomenon, termed evolutionary suicide, has been explicitly modelled by Gyllenberg et al. (2002). They show that under a range of conditions in which habitat patches can become unsuitable, evolutionary suicide can occur.

The type of competition per patch is also predicted to influence dispersal. One of the benefits of dispersal is reduced competition among residents that are kin (Hamilton and May 1977). Inbreeding and population structures that promote kin-groupings should increase those benefits. Given these expectations, interesting predictions can be made about the sex that is expected to disperse under different mating systems (Perrin and Mazalov 2000). The sex that is expected to disperse the most, depends on which sex experiences the most severe form of competition and for what resource. When competition affects both sexes equally, no sex bias in dispersal is expected. In contrast, under polygyny or promiscuity, competition for mates among males exceeds competition for resources among females, thus males should disperse more than females. Male dispersal reduces the chances of females dispersing, because related females are unlikely to become inbred. In many monogamous breeding systems, however, males must defend resources to attract females, so females might be expected to disperse more.

A final set of predictions relates to populations not at equilibrium. Many populations show transient fluxes in geographic range, either because of changing climate or invasions into new habitats, or because they are declining (see Chapter 13). Populations whose ranges are expanding are selected for an increase in dispersal at the expanding range front due to the local appearance of newly suitable patches (Travis and Dytham 2002).

## 6.2   Evidence for dispersal evolution

What evidence supports these predictions? It must be admitted that the theoretical richness of dispersal evolution, while matching easily that on sex ratios, is not yet supported by the same wealth of evidence. The reason is primarily the difficulty of measuring the variables: sex ratios simply require counts of males and females, whereas direct measurement of dispersal rates or distances is extremely problematic. Most studies have used some sort of morphological marker for dispersal ability, such as presence or absence of wings in animals, and presence of dispersive seed structures, such as pappi in plants. Nonetheless, many of the theoretical predictions have received empirical support. The prediction with greatest support is undoubtably the effect of temporal heterogeneity.

Southwood (1962) was one of the first researchers to gather evidence in support of the temporal variability hypothesis. He assembled supporting evidence from the existing literature on insect dispersal, including incidence of migratory habits, and catches in airborne insect traps. His paper was not supported by explicit statistics but are consistent with the hypothesis. For example, most species of British dragonfly that migrate also use temporary water bodies like lakes or ponds for breeding habitat. Of the species from rivers, streams, or bogs, none migrate. Roff (1990) did a similarly comprehensive review, this time supported by statistics, using presence or absence of wings as the marker for dispersal. The proportion of species without wings differed significantly among habitats, with woodlands, deserts, and the ocean surface having particularly high levels of winglessness, which he interpreted in favour of the hypothesis. In a survey of North American grasshoppers and crickets (Orthoptera), he categorized species as either being fully winged, wing dimorphic, or fully wingless. The flightless forms were predominantly found in caves, ant nests, alpine areas, and tundra; again habitats that can be interpreted to have low temporal variability.

In a similar study, Denno *et al.* (1991) surveyed the incidence of the macropterous (fully winged) state in 35 species of planthopping bugs (Delphacidae), and then quantified the persistence of their habitats in relation to the bug lifespan. Again, species have a higher incidence of flight capability in habitats with low persistence times, which happened to be mainly agricultural crops. Within species, it would also make sense for individuals to be able to adjust their dispersal rates according to their individual assessment of local conditions, if such a mechanism exists. Some species indeed display such a plastic dispersive response. For example, the ruderal weed *Crepis sancta*, which grows in the Mediterranean area of France, produces a greater proportion of seeds with dispersive structures under nutrient depletion (Imbert and Ronce 2001) (Figure 6.3).

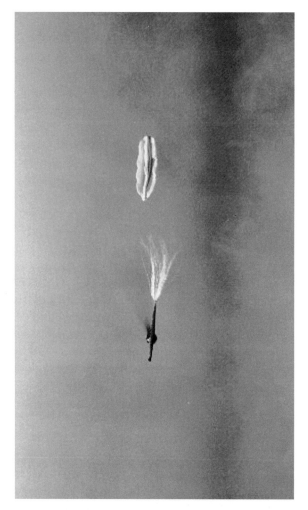

**Fig. 6.3** Seeds of *C. sancta*. The seed at the top (a 'peripheral achene') lacks a parachute and is non-dispersive, while the seed at the bottom (a 'central achene') has a parachute and is dispersive. Under nutrient stress, a greater proportion of dispersive seeds is produced. Both seeds are about 4 mm long. Photo courtesy of Eric Imbert.

So some evidence supports the notion that variation over time in habitat quality has selected for dispersal ability. What about spatial variability? Here the evidence is more indirect because spatial variability is hard to quantify. However, once again we may be able to use rough markers. Oceanic islands are areas whose suitability unambiguously decreases rapidly in space, as the ocean is reached! As mentioned in Chapter 1, in many bird species, incidence on islands is associated with the development of flightlessness (see Figure 13.3). The phenomenon, however, is not restricted to birds; there are many known instances of plant species loosing their seed dispersal structures once island-locked. Cody and Overton (1996) showed rapid reduction of seed dispersal structures in species from the daisy family (Asteraceae) on islands in Vancouver Sound. The evidence from insects (Roff 1990) is more equivocal, since island communities seem to have approximately the same levels of flightlessness as equivalent mainlands (of similar altitude and latitude). However, given that wingless forms are likely to have been under-represented in the initial colonization of the islands, this may still represent a substantial overall reduction in dispersal ability over time (Grant 1998).

Analogous cases might also come from other island-like habitats, whose persistence is assured in the short term but which become rapidly unsuitable over space. One such habitat may be hollow trees, which can persist for many years but which are rather rare in a woodland landscape overall. The beetle *Osmoderma eremita* (Figure 6.4) lives in such hollow trees, and only about 15% of the beetles disperse from the tree they were born in each year (Ranius and Hedin 2001). This makes sense given the kind of landscape they live in: the tree they were born in will likely be suitable for a number of generations, and dispersing individuals may have difficulty finding another suitable tree. Therefore it makes sense for most individuals to stay put, just like in many oceanic island organisms. However, hollow trees are not fully permanent habitats; new ones arise and old ones disappear. Thus individuals will also gain some fitness if a small fraction of their offspring is dispersive.

The contributions of inbreeding and competition have been rather scantily tested, but there is a strong relationship between the sex that disperses and the mating system in birds and mammals, as predicted by theory. In a review

(a)                              (b)

**Fig. 6.4**   The beetle *O. eremita* (a), and its habitat (b), a hollow tree. The tree hollow contains a circular pitfall trap, into which a beetle is about to fall. Trapping studies have shown that only about 15% of these beetles disperse from their native tree, a finding that is consistent with theory of the evolution of dispersal. Photos courtesy of Thomas Ranius.

of studies of dispersal in birds and mammals, Johnson and Gaines (1990) found that polygynous and promiscuous species show male biased dispersal, while in monogamous species either both species disperse or females only.

Finally, evidence is beginning to suggest that expanding populations do indeed have higher dispersal ability towards the edge of their range. During the expansion of the Lodgepole Pine in North America, the seed morphology shifted towards those with higher dispersal propensity (Cwynar and MacDonald 1987). In several British insects that have recently expanded their range due to climate warming, populations at the edge of the range may have evolved increased dispersal relative to older more established populations. The speckled wood butterfly *Pararge aegeria* is one such species. Populations at the range margin have more massive thoraxes, which house the flight musculature and is known to be correlated with flight ability. They also have reduced fecundity, suggesting a fecundity trade-off associated with increased dispersal ability (Hughes *et al.* 2003). There have been no studies on declining populations for comparison, but the studies on expanding populations, while giving comfort for those particular species, give us cause for considerable concern about other species that, despite climate warming, have seen their ranges contract through habitat destruction (Warren *et al.* 2001). It may be that with increasing isolation they begin to behave like island populations and disperse even less than before.

## 6.3    Dormancy and other seed strategies

So dispersal is predicted to increase or diminish in response to a number of selective pressures and there is growing evidence in favour of those predictions. Seed dormancy has a similarly rich theoretical history. Cohen (1966) imagined an annual plant living in a temporally varying environment. Seeds can either germinate or remain dormant to the next year, in which case they suffer a mortality cost. He then asked what proportion of germination maximizes long-term population growth rate (a non-ESS approach). Unsurprisingly, as the probability of total reproductive failure in any year increases, so does the optimal dormant fraction of seeds. Dormancy, like dispersal, is therefore favoured by environments that vary in time. The parallels do not end there: dormancy is also favoured by competition between occupants of a patch, and especially if the occupants are sibs, just like dispersal (e.g. Ellner 1987).

There is evidence to connect temporal heterogeneity with dormancy and competition. Pake and Venable (1996) quantified dormancy in winter annuals of the Sonoran desert, and found that species with higher temporal variation in reproductive success had lower germination fractions. Hyatt and Evans (1998) found a weak but significant negative association between family size and germination fraction in the desert mustard *Lesquerella fendleri*, consistent with the idea that increased levels of sibling competition select for increased levels of dormancy.

There is, however, yet another seed characteristic that can be selected for under the same pressures: seed size. Larger seeds are generally favoured under adverse circumstances for germination, such as competition and unsuitable weather. Therefore plants are actually faced with (at least) three alternative seed strategies to cope with kin competition, crowding, and heterogeneity: dispersal, dormancy, or larger seeds. Venable and Brown (1988), in a wonderful paper that unifies the theory on these three different life history characteristics, have explored whether these alternative strategies actually have different fitness consequences in the face of environmental variability. All three can work as bet hedging strategies to spread risk. All reduce the year-to-year variation in fitness at the expense of reducing average (arithmetic mean) fitness: dormancy increases fitness in unfavourable years, hence reduces fitness variance across years, but at the cost of reduced population growth rate in favourable years and reduced total germination. Dispersal reduces the spatial variance in fitness but at the cost of mortality or fecundity. Seed size lowers the temporal variance in fitness because it improves fitness in unfavourable years, but decreases seed yield in favourable years because large seeds cost more than small seeds. Suppose now that in a variable environment, there is a single optimum balance of the mean and variance in fitness. Increases in the level of one seed characteristic, say dispersal, will demand reductions in one or both of the others. There should therefore be a trade-off between these three variables, one not governed by a constraint but one due to selection.

In their model, Venable and Brown consider what levels of germination rate, seed yield per germinating seed, and the probability of dispersal maximize long-term fitness gain. The model considers five variables and how they affect selection for dispersal, dormancy, or seed size. The results are understandable if one views dispersal as a way of escaping in space, dormancy as escaping in time, and seed size as local endurance (roughing it).

First, consider variation in the number of habitat patches. Increasing this creates selection for increased dispersal, decreased dormancy, and reduction in seed size. The reason is simple: the opportunity to escape in space is increased. As a correlated response, the other two traits decrease. Second, consider decreases in the likelihood of favourable conditions. This always increases seed size, initially decreases dormancy but later increases it again, and initially increases dispersal but later decreases it. Increasingly unfavourable conditions naturally favour endurance strategies. Dispersal is initially favoured if there are only a few unfavourable patches, but less so when nearly the entire habitat is unfavourable. Then conditions favour dormancy instead. Third, localized dispersal creates selection for reduced dispersal and as a result more dormancy and larger seeds, a very intuitive result. Fourth, if patch suitability is similar in space, this selects for decreased dispersal, increased dormancy, and increased seed size. The effectiveness of dispersal as a means for escaping unfavourable conditions is reduced. Finally, if patch suitabilities are similar in time, this decreases dispersal and dormancy but increases seed size. It is obviously not possible to escape in space, so dispersal should not be favoured here.

Hence not only do seed size, seed dormancy, and seed dispersal trade-off against each other, but the optimal balance between them can evolve in response to subtle changes in the nature of the environment. Comparisons between different species, particularly among the well-described British flora, provide excellent evidence that these seed properties trade-off against each other. Rees (1993, 1997) has compared dispersal measures, seed size, and dormancy propensity in several datasets. All three traits are negatively related to each other. Although there have been few explicit tests of the environmental correlates, the theoretical predictions match anecdotal observations on these traits. For example, in many island plants, not only has dispersal decreased but seed size has increased (see Carlquist 1965, 1974). Indeed the largest seed of any plant, the Coco-de-mer, is an endemic of the Seychelles islands (see Edwards *et al.* 2002). These observations fit the predictions about the number of habitat patches. Seed dormancy is reputed to be common in desert plants, where the probability of favourable years (rain) is especially low. In general, however, there is a great opportunity for these predictions to be tested by long-term experiments that either manipulate or at least measure these variables in a quantifiable fashion.

Dispersal and dormancy are both costly traits and this realization set in motion two initially independent research programmes. Both traits are exciting because both are favoured by the same broad sets of environmental variables, and, at least in plants, need to be considered in concert. Dispersal in both plants and animals impinges on numerous other life history traits (seed size, fecundity), and perhaps it is time that both should rightly take their place among the core of life history theory. In addition, dispersal and dormancy evolution provide a clear link between life histories and other aspects of ecology and evolution.

## 6.4   Further reading

Johnson and Gaines (1990) provide a useful overview of classical dispersal evolution theory, and Rees (1997) provides a good introduction to dormancy. Hamilton and May (1977) and Hamilton's own (1996) commentary on this are worthwhile dips.

# 7 Doing adaptive things

Make the most of the best, and the least of the worst.

Robert Louis Stevenson

Most of the traits we have considered so far involve characteristics that differ between populations or species, such as their average sex ratio, their average lifespan, and whether they are sexual or asexual. Individual organisms also possess the ability to modify what they do in response to changes they sense in their environment during their lifetime. In animals, where much of this change involves differences in what they do, we call this behaviour. In plants the differences result from changes in growth, development, and morphology.

Where animal behaviour is considered in the light of adaptation, this is called behavioural ecology. Behavioural ecology actually accounts for the majority of work in evolutionary ecology, largely because many of the conceptual advances that combined evolutionary and ecological thinking have come from considering animal behaviour. There have been numerous reviews, both academic and popular, of the core behavioural ecology material, and it is not my intention to repeat it all here. Instead, I want to show how many of the concepts have a much wider application than just to animal behaviour. Hence, to provide some balance, this chapter will focus on plants. I will cover three key concepts that crop up in other areas of the book: foraging, social evolution, and sexual selection.

Plants display numerous plastic responses to environmental stimuli during their lives that are analogous to animal behaviour (Silvertown and Gordon 1989). These include widespread responses, such as etiolation (lengthening of shoot under shade) and tropisms (growth towards or away from a gradient stimulus), as well as more unique responses, such as sex determination in ferns (females produce a chemical that induces male formation in nearby plants). In addition, behavioural ecology is concerned with very broad features of biological life: reproduction, interactions with relatives, resource acquisition, competition, to name but a few. These are, of course, features of plant as well as animal biology, hence should enlighten both.

## 7.1   Resource acquisition

Let us see how this works. One of the most obvious things that animals and plants have in common is the need for resource acquisition in an environment that is not uniform. In behavioural ecology perhaps the most influential model surrounding this problem has been the marginal value theorem (Charnov 1976). In outline, the marginal value theorem is an optimization model based on economic principles. Imagine an animal foraging in an environment consisting of patches of food of varying quality. After arriving at a patch, how long should the animal continue to forage there, before leaving for the next patch? It should not stay indefinitely in a single patch because there will be diminishing returns. The model calculates the optimum time to stay as that which maximizes the rate of return of energy across all patches. The benefit of staying in a patch is that an animal encounters more food without having to spend time travelling between patches, and the cost is that it has to spend an increasing time within the patch finding the next food item. It is intuitive, and simple to show, that the animal should stay in the patch until its rate of gain of food drops below its average gain in the environment as a whole (i.e. until it becomes more profitable to move on). This will naturally result in spending more time in richer patches. If the travel time between patches is longer, the average gains in the environment as a whole will be reduced, and the animal should stay longer in each patch before moving on (Figure 7.1). These results are intuitive to us, because if we have to travel a long way to the nearest shop, we tend to do all our shopping

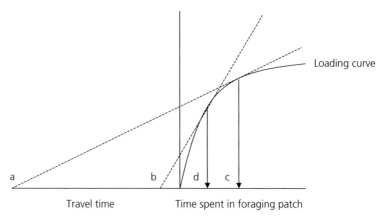

**Fig. 7.1**   The marginal value theorem, following Charnov (1976). The *x*-axis represents time spent travelling (to the left of the *y*-axis) or time at the foraging patch (to the right of the *y*-axis). The *y*-axis shows food gained, described by the loading curve (solid curve), which asymptotes due to patch depletion. The optimal leaving time (vertical arrows) is found by drawing the tangent from the central place to the loading curve, here shown for a long (a) and a short (b) travel time. The optimal leaving time is longer for the former (c) than the latter case (d).

in one go, whereas if the nearest shop is just around the corner, we are likely to do more frequent but shorter visits. A range of animals display these qualitative predictions, making the marginal value theorem a useful heuristic tool.

What about plants? Plants of course are sessile, hence do not forage in quite the same way. Resources tend not to be other organisms (though this does happen in a number of species, see below). Instead, plants use their above-ground modules (leaves, stems, and sexual organs) to capture light, attract pollinators, and to disperse seeds. Below-ground parts capture water and ions. Rather than moving bodily into new foraging patches, plants must direct the growth of new modules into resource rich patches. An explicit use of the marginal value theorem to help understand this problem was by Kelly (1990), who studied host choice in Dodder. Dodder is a rootless, leafless, and non-photosynthetic parasitic plant that coils around the stems of other plants and taps into their vascular system (Figure 7.2). It is essentially just a stem that grows from one host to the next. Dodder individuals can attack a number of plant species over the course of a season, and cover many square metres of ground. Plant species presumably vary in their suitability. Kelly therefore drew parallels between the food patches of the marginal value theorem, and host individuals, and between time spent in a food patch and investment in coiling around a host plant. Dodder plants that maximize their long-term rate of gain in resources will invest more coils around better

**Fig. 7.2**   A shoot of dodder, *Cuscuta europaea*, coiling around the stem of a nettle host, *Urtica dioica*. Photo courtesy of Colleen Kelly.

quality hosts. In a series of transfer experiments, Kelly showed that plant species in which the Dodder invested longer coils gave greater growth per unit coil length, as predicted. In addition, greater growth translated into greater survivorship and fecundity of the Dodder plant.

Although Dodder illustrates that some models of adaptive behaviour can add to understanding of plastic plant responses, parasitic plants are not archetypal of the botanical world. Gleeson and Fry (1997) have made a more representative comparison. They used the marginal value theorem to understand root proliferation in soil patches varying in nutrient concentration. They assumed that each plant had a limited amount of root to invest, soil patches deplete, and plants want to maximize total uptake rate across patches. Here, the optimal investment is where the rate of return per unit root invested is equal across patches. As long as richer patches give consistently higher gains per root than poorer patches, there will be a positive relationship between root proliferation and soil patch quality. Testing their prediction on the grass *Sorghum vulgare*, they found that plants grown in soil patches that differed more in nutrient concentration also showed greater differences in root proliferation in the expected direction. Similar qualitative trends in root growth have been observed in a number of species, and the above-ground parts of plants similarly display growth strategies that enhance their rate of light capture in areas with higher light intensities (de Kroon and Hutchings 1995).

Clonal plants face similar problems at larger scales. These plants consist of plant units, called ramets, which produce lateral extensions (stolons or rhizomes) from which new ramets develop (Figure 7.3): strawberry plants are a familiar example. The environment can vary at the scale of the whole ramet, such that some ramets may develop in uniformly poor patches

(a)                                                    (b)

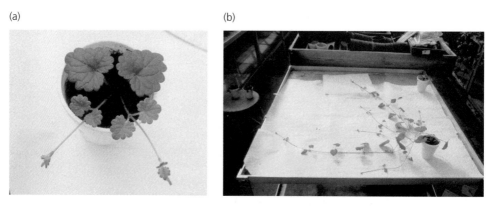

**Fig. 7.3**     Ground ivy, *G. hederacea*, grown in a pot—(a) stolons bearing leaves, with petioles, and (b) a plant 'foraging' with stolons and petioles, across the greenhouse bench. Photos courtesy of Mike Hutchings.

(nutrient poor soil or low light intensity) while others develop in uniformly good patches (nutrient rich soil or high light intensity). Sutherland and Stillman (1988) found that, as predicted by foraging theory, clonal plant stolons or rhizomes are more likely to branch in rich patches, but do not alter their angle of branch. However, Wijesinghe and Hutchings (1996) found that the woodland herb ground ivy (*Glechoma hederacea*) increases allocation to leaves and stolon branching frequency in patches of high light intensity, and lengthens its petioles in patches of low light intensity. Such responses are mainly properties of ramets. They suggest that stolons may not be adapted for optimal placement of ramets, but optimal sampling of the overall habitat. They suggest therefore that clonal plants should be considered as integrated units which can use both ramet and stolon to optimize their use of the habitat. Theory combining these concepts is therefore needed if we are to understand clonal plant proliferation. Whatever the long-term outcome of this suggestion, it is clear that foraging concepts, and particularly those provided by the marginal value theorem have value when applied to plant resource acquisition.

## 7.2   Social evolution

One of the most intriguing features of animals is their behaviour towards conspecifics (social behaviour). This behaviour can vary from cooperation and altruism to competition and lethal combat. Cooperation and altruism, in particular, have been the focus of attention because selfish entities are, by-and-large, expected through natural selection (Chapter 2). However, several mechanisms have been identified by which cooperation and altruism can evolve (see also Chapter 11). The most famous, kin selection, occurs when interactions occur between relatives, and is largely due to work by Hamilton (1963, 1964a, b).

Hamilton derived from **population genetics** a general rule that would predict when a gene for altruism would be favoured: when $br > c$. The three terms of this inequality are respectively, the fitness benefits accrued to the recipient of altruism ($b$), the fitness cost accruing to the altruist ($c$), and Wright's coefficient of relatedness ($r$) which measures the probability that a gene in the altruist is also in the recipient. We have already encountered the idea of relatedness in passing in Chapter 5 and live with these values from day-to-day: for sexual diplo-diploids it is 0.5 between parents and offspring, 0.5 between siblings sharing both parents, 0.25 between grandparent and grandchild (Figure 7.4). The effect of $r$ in Hamilton's rule is that higher values make it more likely that the inequality will be satisfied, hence more likely

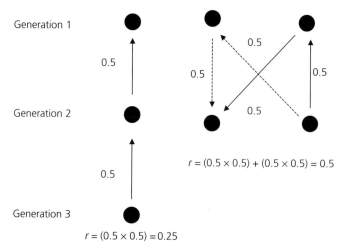

Generation 1

0.5

0.5

0.5

0.5

0.5

Generation 2

0.5

$r = (0.5 \times 0.5) + (0.5 \times 0.5) = 0.5$

0.5

Generation 3

$r = (0.5 \times 0.5) = 0.25$

**Fig. 7.4** Working out coefficients of relatedness. The arrows trace the possible lines of descent of a gene, while numbers indicate the probability of that line of descent under Mendelian inheritance. The relatedness of a grandchild to its grandparent is 0.25 (left), while that of sibs is 0.5 (right).

that altruism will be favoured. This of course is a very intuitive result: we are nearly always nice to ourselves ($r = 1$), we look after our offspring very well ($r = 0.5$), while the concept of the wicked stepmother ($r = 0$) is found in many cultures. Hamilton's rule has found wide support in a number of animal systems, but it is worth remembering that it is only exactly correct under certain conditions, which are not always fulfilled (see Grafen 1984) whereby one may have to resort to basic population genetics.

While Hamilton's rule is widely acknowledged to define limits to altruism, the flip side is that it also defines limits to selfish behaviour. Imagine a situation where the decision is to keep a resource or to donate it to a relative. If the resource has the same value for donor and recipient ($b = c$), then an individual should prefer the resource to belong to itself than another individual, even a sibling or offspring. The lower the relatedness between individuals, the stronger the selfish tendency. Thus, the interests of relatives frequently may not coincide, leading to conflict between family members. Clarence Darrow's famous saying, that 'the first half of our lives is ruined by our parents and the last half by our children' holds some truth for many organisms.

Plants have, in this sense, plenty of opportunity for social interactions between relatives. The application of Hamilton's rule, and especially the concept of relatedness, helps our understanding of many reproductive phenomena. The reproductive biology of angiosperms outlines many opportunities for conflict (see Mock and Parker 1997). Pollen consists of two haploid cells (microspores) produced by a meiotic division encased in a

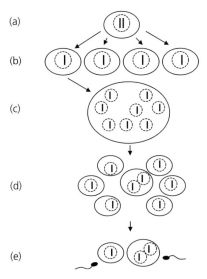

**Fig. 7.5**   The formation of an ovule in angiosperms (thick lines are chromosomes, dotted circles are nuclei, solid circles are cells). A diploid mother cell (a) undergoes meiosis to form four haploid megaspores (b). Three disintegrate, while the surviving functional megaspore undergoes three mitotic divisions producing a single cell with eight haploid nuclei (c). An uneven cytoplasmic division then produces six haploid cells and one diploid cell (d). Once fertilized by the two pollen sperm, the diploid cell becomes the triploid endosperm, and one of the haploid cells the egg nucleus (e).

tough capsule. On arrival at the female stigma one cell grows as a pollen tube, while the other divides by mitosis into two sperm. Eventually the pollen tube reaches the ovary which consists of one or more ovules, which when fertilized will develop into seeds. Like pollen, ovules are rather complicated entities (Figure 7.5). The ovule starts out as a single diploid mother cell which divides by meiosis into four haploid megaspores. Three of the four megaspores disintegrate, leaving a single functional megaspore. This undergoes mitosis three times without cytoplasmic division to give a single cell with eight haploid nuclei. Then a cytoplasmic division occurs leaving six small haploid cells and one central cell with two haploid nuclei. When fertilized by one of the sperm, this cell will become the triploid endosperm that provides nutrition for the developing seed. Meanwhile, one of the haploid cells (the egg nucleus) is fertilized by the other sperm to become an embryo.

The triploid endosperm is a curious phenomenon. The endosperm provisions the seed by drawing on maternal resources. It contains two identical copies of the maternal genome and one paternal. There are at least two reasons why this might have happened. First, it could have arisen to try to counter sexual conflict between mother and father. We will encounter sexual conflict again in a later chapter in a different context, but we have already encountered one of its consequences: genomic imprinting (Chapter 2).

If embryos all have different fathers, then the paternal genome is selected to gather resources for the embryo it has fertilized at the expense of other embryos, to which it is totally unrelated. A mother, however, is related to all embryos equally since she is mother to them all. In short, fathers and mother conflict about apportionment of resources between embryos. Genomic imprinting results from this conflict: it is where the expression of a gene is conditional on the sex that transmits it (Chapter 2). Fathers, for example, should be selected to activate genes that result in resource transmission to the embryo; mothers should be selected to counter this. This is found in maize where the number of paternal and maternal genomes in the endosperm can be manipulated to some extent (Haig and Westoby 1991). Genes on chromosome 10 are only active when paternally derived and result in normal sized kernels if the endosperm genome ratio is normal, small kernels if there is maternal overdose, and when the endosperm is tetraploid with equal maternal and paternal representation, the embryos abort. The latter is possibly a result of active maternal countermeasures. From this example, the doubling of the maternal genome could therefore be a 'trumping' mechanism for the maternal sporophyte to reassert control over the paternal genome.

The alternative view stems from conflict between kin. From an inclusive fitness perspective, the mother is indifferent to which embryo receives resources (but see below), since she is equally related to all of them ($r = 0.5$). The embryo, however, is related to other embryos by between 0.5 (if they all have the same father) and 0.25 (if they all have different fathers) so prefers a skew of resources towards itself. Thus, embryos should be selected to try to gain resources from the mother at the expense of other embryos. There is thus a potential for kin conflict, and hence parent–offspring conflict. At the centre of all this is the endosperm. The relatedness of the endosperm to these players is key. If the ancestral endosperm were a diploid maternal genome, then that would have put the mother in control of resource provisioning, since the endosperm and the maternal interests would be exactly congruent. If the ancestral endosperm was a diploid identical twin of the embryo, that would put the embryo in control. What we have instead is an intermediate situation: a gene in the endosperm is guaranteed to be in the embryo ($r = 1$), but a gene in the endosperm has a two-third chance of being in the sporophyte ($r = 0.67$), and is more related to other embryos than its own embryo is. Hence, things obviously do not go all the mother's way, but she is much better off than if she only had a half share in a diploid endosperm.

There is evidence as well for kin conflict in plants. In *Dalbergia sissoo*, a tropical tree from the pea family, multi-seed pods become single seeded by progressive seed abortion, caused by water-soluble chemicals that diffuse from the focal sibling. This increases the weight of the focal seedling, and the removal of its sibs enables the pod to disperse farther by the wind (Ganeshaiah

(a)                                                                   (b)

**Fig. 7.6**    A flower (a) of *M. guttatus*, and (b) plants growing by a copper-polluted stream at Copperopolis, California. Photos Courtesy of Mark Macnair.

and Uma Shaanker 1988). Sibling conflict of this kind is also probably in maternal interests in many plants. It is characteristic of most flowering plants that there is extravagant overproduction of early reproductive structures in most years. There is strong evidence that mothers are able to select which should stay, and which should go. These include selfed embryos, which are in many plants more likely to be dropped than outcrossed embryos. However, other characteristics may be selected. Amazingly, *Mimulus guttatus* growing in copper-stressed soil can selectively abort offspring which are copper sensitive (Searcy and Macnair 1993) (Figure 7.6). In general there is good evidence that selective abortion improves progeny fitness, that is, that it is adaptive. For example, plants of *Lotus corniculatus* allowed to abort seeds naturally have more fecund offspring than ones which have equivalent numbers hand thinned at random (Stephenson and Winsor 1986).

Even when the fruit has left the parent plant, interactions between the parent and offspring may continue. As Ellner has pointed out (1986), parents may be selected towards making a certain proportion of seeds dormant (Chapter 6). If there is competition between germinating sibs, parents may be selected to produce dormant seeds to reduce sib competition. She values all her offspring equally. Each seed, however, is expected to give priority to itself. This can select for reduced dormancy, giving rise to conflict between parent and offspring over whether seeds should be dormant or not. Rather interestingly, dormancy is commonly caused by a tough seed coat, which is broken down by mechanical or chemical means. The seed coat is a maternally derived tissue, giving her control over seed dormancy.

So the strange reproductive life cycles of plants leave much room for family interactions, through (1) maternal over-production, (2) multiple siring of offspring on a plant, and (3) provisioning of seeds from maternal resources. The concept of kin selection enhances our ability to understand these peculiarities of life.

## 7.3   Sexual selection in plants

This discussion of plant reproduction is convenient for the next subject: sexual selection. Sexual selection was another brainchild of Darwin (1871) who realized that characters that made organisms successful at acquiring a mate might be selected for, and that this can account for some of the extraordinary exaggerated traits of organisms, such as antlers of deer and long tails and colourful plumage of pheasants. The first person to explicitly model female choice as the exaggerating process was Fisher, who imagined runaway selection (Chapter 1). Females would initially prefer some males more than others and these males would then gain greater reproductive success. Over time the preference and the male trait would become coupled and increase in magnitude until countered by some cost of the male trait. A second more recent set of models, collectively known as good gene models, assume that female choice is costly and that therefore it should not be arbitrary. Instead, females should be selected to choose males that provide them or their offspring with some kind of advantage in natural selection. The particular traits females choose should in fact be correlated with some fitness advantage in natural selection.

Are there any plant traits that might have evolved through sexual selection? The obviously exaggerated plant reproductive traits are large showy flowers and long **pistils**, which may have evolved by male–male competition. In the former case, one can imagine male flowers competing with each other for pollinators to remove pollen. The phenomenon of *fleur-du-mâle*, whereby male flowers are overproduced and then aborted, may well be a sexually selected phenomenon: a plant makes itself as showy as possible to attract as many pollinators as possible into the vicinity. For example, many *Agave mckelveyama* flowers are aborted before fruit initiation, and these are always functionally male, and give high nectar rewards (Sutherland 1987). Therefore some flowers, indeed, seem to serve merely to attract pollinators and have no fruiting function. Queller (1983) studied *Asclepias exaltata* and showed that inflorescence size is positively related to pollen removal, while seed set is unaffected. This therefore suggests that the showy trait benefits male function much more than female function.

Despite the possible role of sexual selection in flower evolution, plants in general do not work through the showy media of vision and sound that animals do, so we should expect the effects of sexual selection to be more cryptic. In particular, we might expect evolution of competition between pollen. Pollen grains are known to inhibit each other's growth in some species, and in other species pollen grain size is related to fertilization success. Female choice and male–male competition are likely to interact through pollen competition, as the female tissues provide the arena for pollen competition. In peach trees, the base of the style reduces after pollination and this increases competition between pollen grains by reduction in both resource flow and space (Herrero and Hormaza 1996). Pollen competitiveness also sometimes reflects male quality: pollen tube growth rate in violets increases with the seed production of the donor in violets, and this translates into offspring vigour (Skogsmyr and Lankinen 2000).

The interactions between pollen and pollen, and pollen and sporophyte therefore provide intriguing examples of sexual selection at the boundary of male–male competition and female choice. Much more clearly defined examples of female choice occur in seed abortion. The last section has already shown how this can be selective according to male genotype, and also increase offspring vigour. Once again though, the effects of this choice are relatively cryptic to human eyes.

It is clear that the biological differences between plants and animals sometimes provide us with evolutionary outcomes that are relatively unique to one kingdom or the other. It is equally clear, however, that theory developed for applications in one kingdom can be usefully translated across to the other and enhance understanding. While nobody can deny the success of behavioural ecology, the term itself has limited the application of its findings to different taxa. Evolutionary ecology of course has no taxonomic boundaries and it is on this wider stage that the full implications of theory will be realized.

# 7.4   Further reading

Krebs and Davies (1993) introduce behavioural ecology. Silvertown and Gordon (1989) discuss plant 'behaviour', and Mock and Parker (1997) provide an interesting introduction to family issues in plants. Skogsmyr and Lankinen (2002) review sexual selection in plants.

# 8 Evolution and numbers

So far we have covered how ecology causes anagenetic change, the place of ecology being largely subsumed under the term 'selection'. The present chapter, like Chapter 3, explores more the effect of evolution on ecology, thus closing a causation loop. Specifically, we will investigate the effects of anagenetic change on population dynamics, an ecological characteristic, and ask whether we can better understand population level phenomena by incorporating evolutionary assumptions. Further, we will explore an additional theoretical step in the evolutionary process by using assumptions about population dynamics to help estimate the path of evolutionary change. Thus, evolution influences ecology, which influences evolution again.

There are essentially two situations in which knowledge of adaptation can help predict the dynamics of populations. The first is when organisms display plastic phenotypic strategies (such as behaviours) that are adapted to suit particular environments, and these behaviours affect population ecology in some way. In this scenario no evolution occurs *per se* during the timescale of the study, but the plastic responses are assumed to be the outcome of past selection. The second way is, if evolution itself occurs on ecological timescales, by which I mean roughly the length of a human lifespan. This is more common than it might at first appear. It is not a great leap of understanding to realize that changes to a trait can affect a population's dynamics. That then requires ecologists to think about evolution. However, a further logical step is possible. The fitness of phenotypes may be affected by the ecological interactions between individuals, for example if fitness is density- or frequency-dependent. This may then decide which genotypes spread or disappear. Thus, there may be a case for incorporating assumptions about the effect of a trait on population dynamics merely to work out what the final evolutionary state itself is. The theoretical approach that incorporates these assumptions is known as adaptive dynamics, it is relatively new, and we will examine how it works later in the chapter.

## 8.1    Adaptive decision-making and population dynamics

In the last chapter we discussed a foraging model known as the marginal value theorem (Charnov 1976). The model considers a patchy environment with patches differing in productivity and is asked how a single organism should behave in response to them. We assumed that the final phenotypic state of the organism would be that which maximized fitness, according to the optimization principle. We will now consider a whole population of individuals inhabiting the environment, and assume that the food supply in a patch replenishes so that organisms can inhabit them full-time. How will the individuals distribute themselves between patches that vary in productivity?

This problem was first considered by Fretwell and Lucas (1970). Imagine the first organism to arrive in the habitat. It should, following the optimization principle, settle in the most productive patch, where its food intake will be highest. By foraging, however, the first organism reduces the productivity of the patch for subsequent foragers through competition (Figure 8.1). The second individual to arrive in the habitat therefore faces a slightly different decision because the productivity of the best patch will now appear lower and hence closer to that of the other patches. If it is still higher than all the others, however, the second organism should still settle there. Individuals continue to occupy the most productive patch until its apparent productivity to the subsequent individual equals that of the second most productive patch. At that stage subsequent organisms to arrive should alternately choose the first and then the second patch as one is depleted below the level of the other in turn. If all individuals are equal competitors and they know

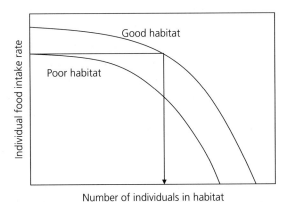

**Fig. 8.1**    The ideal free distribution. The two curves represent food rewards per individual in a rich and poor habitat as the habitat fills up with competitors. The first set of individuals enters the rich habitat, continuing until such time as the individual rewards are the same as in the poor habitat (horizontal line). Thereafter, individuals fill up both habitats such that rewards are the same in both. After Fretwell and Lucas (1970), with permission from Springer Science and Business Media.

both the productivities of the different patches and the distribution of competitors among them, the end distribution, one in which no individual can do better by moving to another patch, is called the ideal free distribution. This is an evolutionary stable strategy, the concept that applies to maximization situations when the fitness of a given strategy depends on the strategies adopted by other individuals in the population (Chapter 6). At the ESS (the ideal free distribution here), the number of individuals per patch is proportional to their productivity, with more individuals in the best patches.

The ideal free distribution model allows us to predict the distribution of organisms among patches and also their fitness in those habitats. The implicit assumptions, that individuals have perfect knowledge of the habitat, of the distribution of conspecifics, and that they can move freely among habitats, are obviously inappropriate for less mobile organisms, but for some, such as migratory birds, they might be fair approximations. The ESS models of habitat selection like the ideal free distribution can then allow us to predict the productivity of individuals in a population in response to changes in density. These 'density-fitness functions' can then be used to predict how the population will respond to changes in the environment, such as removal of some of the available habitat or a decrease in its quality.

Imagine then a migratory bird population that breeds at one location in summer and then spends the winter in another site where it only feeds. Suppose one summer some of the winter-feeding habitat for the species is removed by a new housing development. What will happen to the population? The modelling logic is roughly this (Goss-Custard and Sutherland 1997): the population migrating to the feeding ground for the winter must be packed into a smaller area of habitat. An ESS model such as the ideal free distribution is used to predict the relationship between bird density and *per capita* food intake rate in the new smaller habitat. By making some appropriate assumptions relating food intake to winter mortality, we can describe the relationship between density and mortality for the new smaller habitat. An ESS model can also be used to describe the way that bird breeding-habitat fills up and responds to density, and using some appropriate assumptions, how birth rate relates to density. The equilibrium density of the population will be when the birth rate exactly balances the death rate (Figure 8.2). It is now simple to predict the new equilibrium population size if we know how much habitat has been lost, hence, how much the mortality–density function has been shifted.

Sutherland (1996) has done this for the European oystercatcher (Figure 8.3), a wading bird that winters on western European coasts and breeds throughout

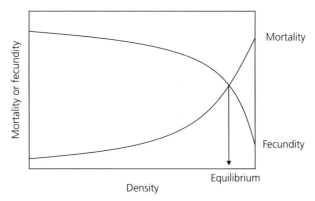

**Fig. 8.2** How to find the equilibrium population density from the per capita mortality and fecundity curves.

**Fig. 8.3** An oystercatcher, *Haematopus ostralagus*, foraging. Photo courtesy of Stephane Moniotte.

the north of the continent. He used ESS models to predict the winter mortality–density function, using data from birds wintering on the Exe estuary in Britain, which harbours around 2,000 birds. From long-term studies of these birds, we know that at high density there is more interference between foraging birds as well as greater food depletion, all of which increases winter mortality. For the slope of the birth rate–density function he used data on a breeding population from the Dutch island of Schriermonnikoog. When density is low on the island, birds can all occupy productive territories by the sea, and the birth rate is high. When density is high, birds are either forced to occupy lower quality territories away from the sea, from which they have to commute to the coast to find food, or they do not breed at all but wait at the coast in the hope of occupying a coastal territory if one becomes vacant. As a result, the *per capita* birth rate is reduced. Consequently, Sutherland was able to predict that if wintering habitat on the Exe estuary were reduced by 1%, the population of oystercatchers would be reduced by 0.69%. The reduction in population size is therefore subproportional to the loss of wintering habitat.

How accurate might the predictions of such 'behavioural-based models' be? In general, we expect them to be better than the major alternative, so-called 'demographic models' which make predictions through direct measurements of the density–mortality and density–birth rate relationships. There are two problems with the demographic approach; one practical and one theoretical. The practical problem is that obtaining accurate data on these relationships can be very difficult. In contrast, studying the behaviour of individuals is much more practical. The theoretical problem is that the new environment, whose effects we want to predict, is very likely to change the density–fitness relationships. For example, reducing the amount of winter feeding-habitat is likely to increase the mortality rate for a given density. The overall behavioural strategy that underlies these changes, however, is likely to remain fixed, therefore by studying behaviour we are studying the mechanistic basis for any change and can detect them.

There is also now some empirical evidence emerging for such advantages of the 'behaviour-based' approach. Stillman *et al.* (2000) have constructed a very detailed behavioural-based model to predict the population responses of oystercatchers on the Exe estuary. The model was constructed using behavioural observations in the late 1970s, when the population was relatively low. Since then there has been an increase in the wintering oystercatcher population. As a result mortality on the estuary has also risen, and the behavioural-based model was able to predict the level of increased mortality relatively well. In contrast, the density–mortality relationship in the late 1970s was a poor predictor of the relationship later on.

We therefore have grounds for optimism in the value of population models based on adaptive behaviour. They have now been applied to a range of bird and some mammal populations, aimed at predicting effects of environment change, such as human disturbance, habitat loss, habitat exploitation, sea level rise, and changing agricultural practice (see Sutherland and Norris 2002). Thus, using adaptive models to predict how organisms will change their behaviour in response to changes in their environment can help us to predict the consequences of those changes for the population biology. This assumes that all the evolution has happened in the past leaving the organism with a plastic adaptive response to the environment. Another possibility exists however, that organisms will evolve to changing circumstances on timescales that are relevant to population ecologists.

## 8.2 Evolution in ecological time

Scientists now agree that evolution can frequently occur over timescales that matter to ecologists, such as a few decades or less. There are few more graphic examples of rapid evolution than of the emergence of new human diseases. In the case of human immunodeficiency virus (HIV), the disease it causes (acquired immune deficiency syndrome or AIDS) was first recognized in 1981. The virus responsible (Figure 8.4) was characterized in 1983 as a retrovirus with similarities to primate lentiviruses, also known as simian immunodeficiency viruses (SIVs). SIVs are widespread among African apes and monkeys, and phylogenetic comparisons of HIV and SIV sequences suggest, with a large margin for error, a likely origin of the current epidemic in the middle decades of the twentieth century (Hahn *et al.* 2000). Most interestingly, SIVs are not known to cause disease in their hosts, whereas HIVs are, as far as is currently known, 100% fatal. It seems therefore that HIVs have evolved to become more virulent over the timescale of only decades. That, however, is nothing compared to what goes on within a single host.

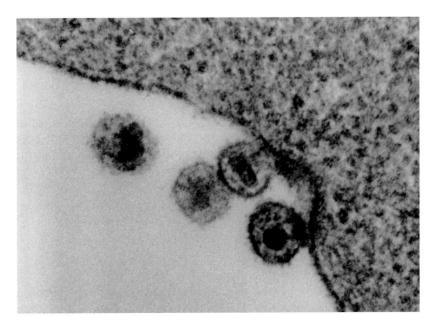

**Fig. 8.4**  Particles of HIV infecting a human tissue-culture cell. The HIV particles are about 100 nm across. Photo courtesy of the Ripple Electron Microscope Facility, Dartmouth College, New Hampshire.

When transmitted to an uninfected human, the virus undergoes rapid replication in the so-called 'primary infection' stage. Replication occurs largely in white blood cells, though other tissues can be used. The immune system of the host then rapidly responds and reduces virus to very low levels. At this stage the host can be recognized as sero-positive. During the next, latent, stage of infection, diversity of the virus slowly increases. New strains of the virus arise because of the low fidelity of the enzyme reverse-transcriptase, which converts the RNA virus genome into DNA that can integrate with the host genome. The rate of **substitution** in the virus genome is roughly one million times that of human genes. Typically $10^{10}$ virus particles can be produced per victim per day. After a variable number of years of viral replication and gradually increasing viral diversity, depletion of **T-lymphocytes** is such that the virus replicates rapidly, and the virus load becomes dominated by fast-replicating strains (Nowak *et al.* 1991). AIDS symptoms occur, and the victim dies of some secondary infection that would not normally be fatal.

We have observations suggesting evolution on two scales: an increase in the virulence of HIVs relative to SIVs and a within-host evolution towards increasing virulence towards the latter stages of infection. Has natural selection been responsible for these trends? It is likely in both cases. For example, resistance to individual antiviral drugs, such as AZT, occurs typically in six months, and similar mutations characterize the resistant strains. This strongly suggests selective evolution of the virus. What about the increase in HIV virulence relative to SIV virulence? Both viruses have high replication rates, mutation rates, and viral loads. These then do not, on their own, seem to engender high virulence. However, the virus envelope protein of SIVs is very conserved, but rapidly evolves in HIVs. In addition SIVs only stimulate a weak immune response from their hosts, while HIVs stimulate a strong immune response.

This suggests that SIVs experience stabilizing selection, while HIVs are positively selected by the immune response. When a strong immune response occurs, genetic variation is created by strains attempting to evade it, that eventually leads to virulence and AIDS. For SIVs a weaker immune response creates reduced selection on the virus, creating less viral competition and less virulence. In summary, the difference in virulence between HIV and SIV seems to have been driven by the extent of arms race between virus and immune response (Holmes 2001). Under this scenario pressures that select for virulence of HIV relative to SIV are the same as those that select for virulence during HIV infections.

There seems little doubt therefore, that HIV can and has rapidly evolved virulence. It seems obvious, especially in the context of a disease like AIDS, that there will be population consequences of short-term evolution. How in general can we predict such consequences? A taste for this comes from recent work on the evolution of exploited fish stocks. Unlike in HIV, a convincing case for short-term evolution in fish stocks has taken time to accumulate. In contrast to HIV, however, we are much better equipped to predict the population consequences because of the presence of high quality data on relevant parameters.

## 8.3 Consequences of short-term evolution

Through harvesting, humans can impose impressive selective forces on populations. The North East Arctic cod *Gadus morhua* (Figure 8.5) has experienced a large reduction in age at maturity from 9 to 6 years in the last 50 years, a change that is typical of other exploited stocks of marine fish in the North Atlantic (Law 2000). One likely reason for this has been the change from fishing only at the spawning grounds off the coast of Norway to fishing largely in the Barents Sea, which is the feeding grounds of the fish. This change in fisheries practice was brought about by the introduction of motorized deep-sea trawlers in the 1920s. Currently almost 40% of the stock

**Fig. 8.5**    A cod, *G. morhua*, on the slab. Fishing activity has led to a decrease in the age of maturity and consequent reduction in yields. Photo courtesy of Gerd-Peter Zauke.

is removed annually from the Barents Sea. Over the course of several years, the cod have very low chance of maturing successfully. Genetic variants that escape the feeding ground to spawn earlier have a much better chance of reproducing. In contrast, when the fish were caught only at the spawning grounds, mortality would have been lower at the feeding grounds, creating selection for increased age at maturity.

Can these changes in fishing mortality exert effects on the fish population? There is some evidence that it can: for example, Conover and Munch (2002) studied the Atlantic silverside (see also Chapter 5). Removing smaller individuals gave an increased yield over selecting larger individuals. There were two reasons: first, when small fish were harvested the adults grew larger, giving greater reproductive potential. Second, removing smaller individuals selected for faster growth as the fish then spend less time in the more vulnerable stages. Such studies show the potential for harvesting to cause evolutionary changes that affect yield. They do not, however, easily suggest what the consequences have been in cod, where the regime is not purely a size selective one and the evolutionary changes have been in age at maturation. For this a specific cod model is needed.

The intuitive effects of reduction in age at maturity are that the yield of fish will decline: the fish mature earlier, hence are putting energy into reproduction instead of growth at an earlier age. Law and Grey (1989) developed the notion of the evolutionarily stable optimal harvesting strategy (ESOHS) to help quantify the effects on yield. At the ESOHS, fishers adopt a strategy that maximizes yield after evolution has reached its final state. They showed that the ESOHS for the North East Arctic Cod was the traditional practice of fishing only at the spawning grounds, and that fishing at both feeding and spawning grounds led to a reduction in long-term yield of up to two fold. One problem, for fishermen, of following the ESOHS is that they may only gain the benefits of maximizing yield after waiting for evolution to come to rest. In the meantime, it is possible that they may have to pursue a suboptimal strategy from the perspective of maximizing yield. Instead it would be sensible for fishers, at least in the short term, to pick the strategy that maximizes sustainable yield at each point in time. In response to this, the fish would experience a slightly different selective pressure, and evolve to a new state, after which the fishermen would perhaps need to change their optimal harvest strategy. This situation is thus, as in HIV infections, a co-evolutionary circuit, this time consisting of evolutionary changes in fish, followed by changes in harvesting strategy by the fishermen.

Heino (1998) has modelled these co-evolutionary dynamics for cod. He was interested to know what particular type of harvesting strategy maximized sustainable yield and whether the co-evolutionary dynamics also lead to an asymptotic fishing strategy that is the same as the ESOHS.

The technique he used to determine the evolution of the fish is known as adaptive dynamics (Dieckmann 1997; Waxman and Gavrilets 2005, see below). Like Law and Grey, he found that fishing in both the spawning and feeding grounds led to a reduction in yield relative to fishing in only the spawning grounds. However, the co-evolutionary dynamics frequently led to the ESOHS even if it was not pursued in the first place.

## 8.4   Adaptive dynamics

The techniques used in Heino's model are interesting. In a normal ESS approach there are no explicit assumptions about population dynamics made in determining the final evolutionary state. If, however, we explicitly want to predict the effect of evolution on population dynamics, we are more likely to have to face up to this challenge. Adaptive dynamics adds in this interaction. In brief, the fitness of new mutants is assessed in terms of their population dynamic interaction with residents. If at the end of that interaction, the mutant has a higher fitness than the resident, the mutant becomes the resident strategy and the resident is displaced. This continues until a so-called 'singularity' is reached (Figure 8.6). Much of the time a singularity may be a fitness maximum (and ESS), thus representing a single resting state. However, in some circumstances the singularity may, counter-intuitively, be a fitness minimum. This means that evolution converges to a point that is less fit than all surrounding strategies. This is surprising to evolutionary biologists who are used to the analogy of evolution as climbing an adaptive landscape, a view that is consistent with the notion of an ESS. However, this

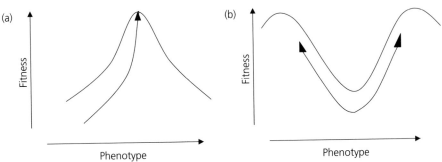

**Fig. 8.6**   Adaptive dynamics leading to evolutionary branching. A lineage normally moves (arrow) over an adaptive evolutionary landscape by climbing to areas of higher fitness (a), eventually reaching an adaptive peak which is fitter than all other local phenotypes. However, when the adaptive peak is reached, the dynamical interactions between the resident genotypes and mutants may allow any alternative new mutant to invade, which means that the resident genotype is now at a fitness minimum (b). Under certain conditions, two divergent phenotypes can then coexist, the start of a polymorphism or even speciation.

intuitive notion assumes that the landscape is fixed, at least during the timescales involved in any one study. Adaptive dynamics changes all that. Invasion of new mutants is assessed in relation to the ecological dynamics of mutants and residents, at each step of the adaptive walk. The dynamics of evolution change the ecological dynamics at each step, and the changing ecological dynamics change the local fitness landscape because fitness can be density- or frequency-dependent. Thus, it is perfectly possible for evolution to proceed towards a fitness minimum because the ecological interaction that determines evolution towards that point is different to the interactions that occur once the point is attained. After climbing to what seems like a peak, sometimes the peak sinks and becomes a trough.

What happens in the neighbourhood of a fitness minimum is particularly interesting, for then any alternative mutant strategy is fitter than the resident: in other words, selection is disruptive. This can lead either to a coexisting polymorphism, or even speciation in sympatry. This so-called 'evolutionary branching' can occur repeatedly. Thus, taking into account the population dynamic interactions involved during evolution can lead to very different predictions about the asymptotic state of evolution, which are not only intuitively appealing but seem to offer explanations for observable phenomena. For example, polymorphic strategies are commonly seen in traits that relate to population level phenomena, such as body size, dispersal rate, dormancy rate, and so on. In addition, the degree of evolutionary branching has obvious implications for speciation, and has been implicated as a major source of sympatric speciation (Dieckmann and Doebeli 1999).

The justification for adaptive dynamics at present is purely logical; it is not yet known to increase predictability. For an evolutionary ecologist, however, there is excitement in exploring the interaction between evolution and population ecology. The perceived need in medical fields, caused by the evolution of disease, has led to the notion of Darwinian medicine (medicine with evolution in mind), and in fishing circles to Darwinian fisheries (fishing with evolution in mind). The next few years should unravel how many areas of population ecology Darwin's name will attach to.

## 8.5   Further reading

The ideal free distribution is introduced by Krebs and Davies (1993). Goss-Custard and Sutherland (1997) and Sutherland and Norris (2002) introduce behaviour-based models. Holmes (2001) discusses HIV evolution. Law (2000) introduces evolution of fish stocks, and Waxman and Gavrilets (2005) and the many commentries that follow it discuss adaptive dynamics.

# 9 A world of specialists

Specialists are people who always repeat the same mistakes.

Walter Gropius

Imagine that most typical of ecological entities, a community of coexisting species. The community may be characterized by variables, such as the number of species and the number or type of interactions. The next several chapters will examine how evolution affects these fundamental properties of communities. This chapter specifically examines the evolution of the breadth of use of resources that govern the size of an organism's ecological niche.

Some species are generalist, having in some sense a broad niche (large range of environmental tolerance, broad diet, etc.). Most species, however, are probably specialists in at least one sense. This is curious from an evolutionary perspective; we can imagine that a wide ecological niche would be extremely beneficial from a fitness perspective, allowing an organism to survive unexpected changes in its environment. The problem then is why so many species apparently forsake such benefits, leading to a great subdivision of resources and environments among species. The widespread evolution of ecological specialization is at least a partial answer to the question 'why are there so many different kinds of animals [and plants]' (Hutchinson 1959).

It is conceptually useful to distinguish two types of niche; the 'fundamental' niche was defined by Hutchinson as the set of environments in which a species could maintain a positive growth rate (i.e. where it could in theory persist). The smaller, 'realized' niche is the subset of the fundamental niche actually occupied in nature. The fundamental niche is governed by an organisms' adaptation to the environment in terms of morphology and physiology. Some of the fundamental niche may then not be used, either due to constraints, such as dispersal ability, or because the organism displays plastic behavioural rejection of some habitats.

Understanding the evolution of the realized niche therefore demands that we (1) understand the extent of the fundamental niche through genetic variation for the preference and exploitation of different numbers of environments, and (2) understand adaptive decision-making (see Chapter 7) that narrows the range of environments actually utilized. We will first explore the very rich body of theory on these subjects before proceeding to the evidence.

## 9.1 Evolution of performance under trade-offs

The first attempt to model the evolution of the fundamental niche was that of Levins (1968). He devised a model that should seem familiar, for in structure it is very similar to the model of the evolution of co-sexuality by Charnov *et al.* (1976) (Chapter 5). Imagine an organism faced with two habitats, A and B, of differing frequency. Its fitness in the two habitats is defined by a trade-off, such that when its fitness in A is high, its fitness in B will be lower, and vice versa, a situation known as antagonistic pleiotropy. A severe trade-off is of a concave shape, but a convex trade-off is less severe, such that at best an organism which is very fit in B will also be pretty fit in A (Figure 9.1). Now suppose that the organism wants to maximize its fitness in the environment as a whole: what point on the fitness curve is best for it? If an organism chooses to maximize fitness in a single habitat, at the expense of fitness in other habitats, it will by definition be a specialist. If fitness is approximately the same in all habitats, it will be a generalist. The basic

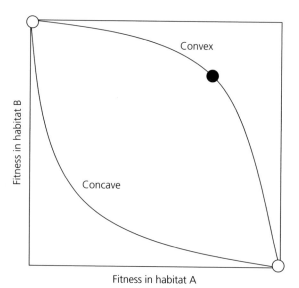

**Fig. 9.1**    Levin's (1968) model of the evolution of specialization. Fitness in two habitats may be described by a trade-off, which may be convex or concave. A convex trade-off leads to the evolution of a generalist, which can use both habitats quite well (black circle), because its total fitness from both habitats is greatest then. A concave trade-off leads to specialization on the most common habitat (white circles), because any alternative generalist would achieve reduced total fitness.

solutions are pretty intuitive: if the trade-off is concave then organisms evolve to maximize their fitness in one habitat, whichever is the most common. If the trade-off is convex, such that a generalist is almost as fit in both habitats as a specialist, then as long as both habitats are reasonably common, generalists will evolve.

According to Levin's model, the evolution of specialism is thus predicted to rely on the presence of a fitness trade-off, the more severe the better, and will also depend on the frequency of alternative habitats, with increasing bias in frequency of one habitat tending to lead to specialism in that habitat. Of course, this model just considers the fundamental niche, and is not explicitly genetic.

## 9.2   Other single species models

Most other models considering the evolution of the fundamental niche have been explicitly genetic (reviewed in Hedrick 1986; Futuyma and Moreno 1988). In general, they consider quite simple genetics, such as one **locus** and two **alleles**, each giving higher fitness in one habitat than the other. They aim in general to predict under what circumstances the population will become fixed for one allele (a specialist population) or retain both (a generalist population), the former case being a 'monomorphism', and the latter a 'polymorphism'. Of course for those interested in specialization, the circumstances favouring monomorphism are the subject of interest. The same models, however, are interesting to another group of scientists: those interested in speciation. They are interested in the conditions that favour polymorphism (Chapter 1). The reason is that once a polymorphism has arisen, with some individuals in a population favouring one habitat and others favouring another, that might eventually lead to the generalist species splitting into two specialist species (Maynard Smith 1966). We will consider a case where speciation is likely to occur in this chapter, and another where speciation already has occurred in Chapter 12.

The predictions of the one locus genetic models are also quite intuitive: if there is temporal turnover in the habitats, a stable polymorphism can arise. This is most people's intuitive reasoning for the advantages of generalism: that it allows an organism to survive changes to its habitat. If, however, there is spatial variation in the distribution of habitats, which is normal, then whether a polymorphism evolves depends on how fitness is affected by population density. If the density of the entire population affects fitness, regardless of which habitat they are in (known as 'hard' selection), specialism is the normal outcome. If, however, it is the density of individuals in each habitat patch that affects fitness (known as 'soft' selection), then a polymorphism can be stable. The reason is again intuitive, since the fitness of each genotype will be frequency dependent, favouring individuals that exploit unoccupied habitats, just as in the ideal-free-distribution model (Chapter 8).

Trade-offs in performance (or preference) in the different habitats are central to all the above models: they assume that a 'jack-of-all-trades is master of none'. Recently, however, attention has turned to models that do not rely on this assumption. The stimulus, as we shall see later, comes from the rather weak empirical support for the existence of fitness trade-offs. For example, Whitlock (1996) imagined competition between a specialist and a generalist, whose fitness was initially the same in the specialist habitat. However, because it experiences only one habitat, the specialist would more rapidly accumulate alleles beneficial in that habitat due to increased strength of selection. Conversely there would be lower accumulation of deleterious alleles. Thus, the specialist's fitness rises faster than the generalist's in that habitat, such that it becomes the superior competitor over time. Specialization thus enhances the speed of adaptation and reduces genetic load: the same long-term advantages invoked for the maintenance of sex (Chapter 2). A number of other models have since explored specialization without trade-offs (reviewed in Futuyma 2001). All differ from the trade-off models in concentrating more on the constraints imposed by the evolutionary process itself.

## 9.3 Niche evolution under competition

Whitlock's model invokes competition between species. A suite of other models have investigated how competition affects niche evolution (e.g. Slatkin 1980; Taper and Case 1985). In many of these models, a species is assumed to display a character that affects its niche use, and that this is under genetic control. The extent of competition between individuals depends on the overlap in niche use. The models run in two phases: first competition is applied and the fitness of the different phenotypes is calculated. Then **quantitative genetics** is applied to calculate the new phenotypes after reproduction has taken place. A carrying capacity is also specified for each environment that determines how the level of competition relates to the number of individuals using that part of the niche.

When only one species is present, it evolves so that its niche gives the highest **carrying capacity** under intraspecific competition (Figure 9.2). When two species are present, displacement of the trait occurs through interspecific competition. Two general outcomes are possible: character divergence and character convergence. Character divergence occurs when the mean phenotypes of the species move away from each other, because the

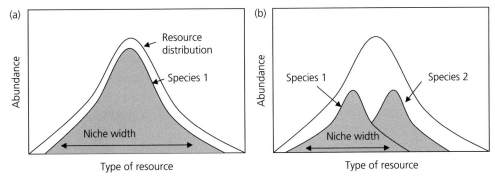

**Fig. 9.2** Reduction of niche width by interspecific competition. (a) When just a single species uses a resource, the average niche position is determined by the most abundant resource type. b) A competitor species will tend to displace the resident, reducing the resident's niche width.

fitness of the phenotypes that overlap is reduced. However, more rarely character convergence can occur. Two circumstances that favour this are (1) when one resource is essential for both species, such that it pays to dominate that resource, and (2) if one resource is particularly abundant while others are limiting.

Thus, competition between species can lead to character divergence among sympatric species, and subsequent specialization, or convergence depending on the circumstances. Of course, there is no reason why competition need be the only interspecific interaction that can affect niche evolution; organisms might, for example, become specialized because their fitness is reduced in some habitats via predation or parasitism. Historically though, such interactions have yet to receive their due share of theoretical attention.

## 9.4 Reduction of the realized niche by decision-making.

Once the genetics of the fundamental niche have been brought to equilibrium, another set of processes can reduce the actual set of resources used. These include individual decision-making, which, though it acts within the constraints of genetics, and hence cannot extend the fundamental niche, can reduce it. In general, the solutions to the optimal decisions are to be found within the realms of behavioural ecology theory (Chapter 7). Foraging models that consider the optimal use of resources are particularly appropriate, and we will consider one more such model here, the optimal diet model, which evolved from work by MacArthur and Pianka (1966). This, like the marginal value theorem, is a rate maximization model, and assumes simply that two resources are available at different frequency in the environment,

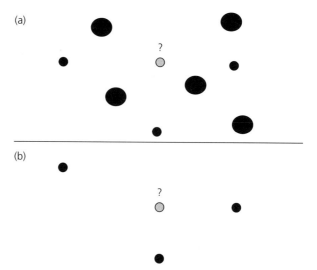

**Fig. 9.3**  Optimal diet selection. An individual arrives at a low value resource (grey circle). Should it use it or move on in the hope of finding a better resource (large circles)? When high value resources are abundant, it is best to move on, because of the opportunities that are lost by staying (a), but when good resources are scarce, it loses nothing by staying (b).

which give different fitness returns (as a result of the outcome of the evolution of the fundamental niche). The resources have different handling times (times taken to exploit them). On encountering the lower quality resource, should an organism accept it or wait until it finds the next higher quality resource? The solution is simple: it should accept it as long as if when doing so it does not gain energy at a lower rate than if it searched for and used the higher quality resource. Therefore, if the better resource is rare or has a larger handling time, or is not much better than the worse resource, all these favour acceptance of the worse resource (Figure 9.3). However, if the best quality resource is far superior, common, or takes little time to handle, then selection favours rejection of the lower quality resource and hence reduction of the realized niche.

Models based on such 'time limited dispersers' have also been written for egg-laying taxa and those selecting habitats to develop in, which for some taxa, such as most herbivorous insects, is the same thing (see Jaenike 1990; Mayhew 1997). The general predictions of these and similar models are that specialization (reduction of the realized niche) is favoured by (1) abundance of the high quality resource; (2) low density of competitors; (3) low variability in resource abundance; (4) longevity; and (5) high egg loads. Unsurprisingly these predictions are very similar to those that predict a specialized fundamental niche, although the variables explored do not always overlap. How do all these predictions stand up to data?

# 9.5   Evidence for assumptions and predictions

Most of the evolutionary models for fundamental niche evolution assume antagonistic pleiotropy, in other words negative genetic correlations between fitness in different habitats. Some negative fitness trade-offs have been shown, and these undoubtedly enhance the process of specialization. For example, pea aphids with a genetic propensity for high fecundity on clover have a genetic propensity for low fecundity on alfalfa, and vice versa (Hawthorne and Via 2001). Rather excitingly, performance on those plants is also correlated with preference for those plants, and could represent an incipient stage in speciation. In general, however, most genetic correlations are not significantly negative and most are close to zero, indicating that high fitness in one environment is relatively independent of fitness in other environments.

There are numerous potential reasons why this result may be rather artefactual, to do with the difficulties of conducting flawless and powerful experiments to measure genetic correlations. On the other hand, the fact that most genetic correlations are near zero, rather than positive, implies that different genotypes favour different environments, rather than a single genotype being best in all environments. The former is needed for the alternative 'neutral' models. In addition, deleterious mutations with habitat-specific effects are known in *Drosophila*, which is an assumption of many of the neutral models (Kawecki 1994).

There is plenty of evidence that variation in resource abundance can favour specialization. Several examples of rapid niche evolution have occurred as introduced species have invaded new habitats and become abundant (Thompson 1998). In most cases this involves evolution of preference for the new habitat. One of these species is the Edith's Checkerspot butterfly *Euphydryas editha* (Figure 9.4), which has been the subject of a long-term study in and around the state of California (Singer *et al.* 1993). Two populations, 'Rabbit' and 'Schneider' have shown long-term changes in preference to lay eggs on novel host plants that have become more abundant in these habitats due to human activity. In both cases the preference differences were heritable, and at Schneider this preference was also associated with elevated growth rates on the preferred host (Singer *et al.* 1988). In addition to such genetic changes, which represent changes in the fundamental niche of a population, there is also plenty of evidence from this and other phytophagous insects that selection of host plants on which performance is poor depends on the abundance of hosts on which performance is high (Mayhew 1997).

Recently, Prinzing (2003) has shown that arthropod species living on tree trunks in Germany are more likely to occupy a large number of microhabitats if they have long generation times, move faster, and spent more of their life

**Fig. 9.4**   Edith's Checkerspot butterflies, *Euphydryas editha*, mating. The butterfly has rapidly evolved to use new host plants as the abundance of hosts in the environment has shifted. Photo courtesy of Camille Parmesan.

on the trunks, giving them better search capabilities and opportunities. The trends might have come about via evolution of the fundamental niche or from plastic behaviour: both sets of theory make the same prediction and only genetic or behavioural studies could separate the two possibilities. Similar correlations have been found between the variability of a habitat and host plant range across herbivorous insect species (Brown and Southwood 1983).

The evidence that interspecific competition causes niche evolution comes from studies of well-defined morphological characters that are functionally linked to resource use—such as the beaks of Darwin's finches (see Figure 12.4). The majority of cases (there are at least 75) cite as evidence that the characters are more divergent when species are in sympatry than when they are allopatric. However, in most cases the case for character displacement by competition is far from complete. Perhaps the very best evidence comes from experimental work by Schluter (2001) on body size and shape of three-spined sticklebacks. In lakes where one species exists, the species is generalized and feeds both in the water column and on the lake bed. When two species coexist, however, they specialize with a thin species in the water column, and a larger rounded species on the lake bed (Figure 9.5). In an elegant series of experiments in artificial ponds, Schluter was able to show selection operating in the predicted direction when an intermediate form was made to coexist with specialists of either type: if it faced benthic competition, the more benthic-like of the intermediate

**Fig. 9.5**    Specialization under competition. The picture shows sticklebacks, *Gasterosteus* spp., from Enos lake on Vancouver Island, British Columbia, Canada. From top to bottom are shown: **limnetic** male, limnetic female, benthic female, benthic male. The benthic forms are larger and have deeper bodies than the limnetic forms. Photo courtesy of Dolph Schluter.

phenotypes had reduced success, thus selection favoured evolution towards a more pelagic existence, and the opposite occurred when a pelagic competitor was introduced. This disruptive selection is a likely mechanism of speciation (see Chapter 12).

In general, the list of character displacement examples, although admittedly preliminary, is remarkable for the absence of character convergence, for the fact that the species involved are closely related, and for the fact that carnivores are very well represented. These may, of course, be artefacts of observer and publication bias, but they may also have a biological basis, a possibility that is intriguing. Can other interspecific interactions cause specialization? There are some intriguing pieces of evidence. There is a cross-species association between host plant range and sequestration of nasty chemicals which serve as defences against predators. This is consistent with predation-driven selection of specialization (Dyer 1995).

There is little evidence that intraspecific competition within habitats can widen fundamental niche use, but certainly the realized niche tends to be greater when densities are high in a large number of species. For example, *Pemphigus* aphids preferentially occupy leaves of their host plants with a good vascular supply which they defend despotically. Subsequent aphids are faced with having to occupy less productive locations on the plant, and at higher densities the range of microhabitats used increases (Whitham 1980). Similar observations could be quoted for almost any territorial species.

In general then, a number of extrinsic ecological and intrinsic biological forces are both predicted to favour restriction of the fundamental and realized niches, and theory and evidence are coming together. It cannot be claimed that we have a good understanding of the relative importance of these forces yet, but a series of hypotheses with some support bodes well for the future.

## 9.6    Evolutionary trends in specialization

Given that we can hypothesize circumstances that might favour specialization over generalization and vice versa, how might the frequency of specialization and generalism vary over evolutionary time? There are three commonly cited hypotheses. First, specialists may be more prone to extinction than generalists. Second, specialists may be more prone to speciation than generalists. Third, lineages may change anagenetically from generalist to specialist more frequently than the reverse. The first hypothesis is intuitive. A generalist seems more likely to maintain a positive rate of increase in the face of environmental change, and a species with a large geographic range or population size would seem less likely to be prone to extinction causing events. The prediction is supported by **metapopulation** models (see Chapter 13), where the greater the density of habitable patches, the less likely the population will go extinct. The second hypothesis is motivated by the observation that a large number of species, especially those in adaptive radiations are ecologically specialized.

The third hypothesis, an anagenetic trend to specialization, was hypothesized by Simpson (1953) to be the dominant trend in adaptive radiation. It could come about for a number of reasons: resource partitioning through competition, adaptation might be quicker through specialists than generalists (see above); specialization might allow speciation through host switching (see above); specialization might canalize evolution through suites of co-adapted traits acting as constraints (Schluter 2000).

Does specialization tend to increase the rate of extinction? Here the evidence strongly supports the hypothesis. Species persistence times in the fossil record are higher for generalist clades in **crinoid** worms, marine **gastropods**, marine **bivalves**, shrews, and antelopes. In recent (anthropogenic) extinctions, there is a tendency for increased extinction risk with decreasing geographic range in primates, carnivores, and birds, and increased extinction risk with increasing dietary specialization in hoverflies, reptiles, birds, and primates (Purvis *et al.* 2000), and with habitat specialization in a number of taxa (Fisher and Owens 2004). Does specialization tend to increase the rate of cladogenesis? Although there is some evidence consistent with this observation, there is also potential counter-evidence. For example, Owens *et al.* (1999) found that bird species richness is positively correlated with dietary and habitat generalism. This, however, might be due to a lower extinction rate or a higher speciation rate, or both.

As regards transitions, trends are seen across phylogenies for increases in specialization over time, increases in generalization and for variable transition rates between the two. Schluter (2000) in a survey of 20 plant, invertebrate, and vertebrate phylogenies found no overall tendency for the ancestor of the radiation to be a specialist rather than a generalist or vice versa. Some taxa, such as the Galapagos finches, had a generalist ancestor, while others, such as the Hawaiian silverswords, had a specialized ancestor. Only in one group, the genus *Aquilegia*, was there a tendency from a generalist to specialist state. In the other groups there was no detectable trend.

The evidence to date suggests that generalization reduces extinction rates, that degree of specialization variably affects speciation rates, and that transitions between specialists and generalists are also variable. The next chapter asks what all these species are specialized at doing, and how that evolves.

## 9.7   Further reading

Futuyma (2001) and Schluter (2001) give introductions to single species and multi-species questions. Futuyma and Moreno (1988) is older but still useful. Mayhew (1997) reviews realized niche models in the context of phytophagous insects. Schluter (2000) reviews macroevolutionary trends.

# 10  The good, the bad, and the commensal

> The web of our life is of a mingled yarn, good and ill together.
>
> William Shakespeare

The present chapter deals with another feature of interactions between species: the degree of antagonism. Interactions between two species can be classified according to the fitness effects of the interaction on each species. Most people's understanding of ecological communities is that they are full of exploiter–victim relationships, where one species benefits at the expense of another (+,−). Predation, herbivory, and parasitism are examples, and are of course ubiquitous in communities: they are the building blocks of food webs. Competition involves negative effects on both parties (−,−) and is also common between members of the same trophic level. Again, such relationships are to be expected in a Darwinian world where individual's primary concern is its own reproductive success, rather than that of another individual of a different species. There are also more nearly neutral relationships, such as commensalisms (+,0) and amensalism (−,0). In these one species is affected by the interaction, but the other is not affected. Commensalism is common where one species incidentally benefits from another species' activity: humans have their own community of such species that includes house sparrows, house mice, house spiders, and most garden weeds, which have little effect on human fitness most of the time. Amensalism is found when one species consistently excludes another (asymmetric competition), as found in many plant and insect communities.

Finally, there are mutualistic relationships (+,+) where both parties benefit from each other. Mutualistic relationships of this kind are ubiquitous and of great ecological and evolutionary significance (Chapters 2 and 3). They include plants and their pollinators; plants and their fruit and seed dispersers; plants and their **mycorrhizal fungi**; corals and their symbiotic algae; lichens (symbioses between fungi and algae); **leguminous plants** and **rhizobia** bacteria; and many animal species with nutritional or digestive symbioses with microbes; including cows, termites, aphids, and humans. Put simply, these mutualistic relationships are essential components of most modern communities.

One striking fact that has emerged from the study of species interactions is that they must sometimes evolve pretty rapidly: different populations of the same species can be either mutualistic or parasitic, and closely related species likewise. In contrast, other relationships are apparently ancient and stable. What conditions favour such changes or else help maintain the status quo? Our understanding of these questions has particularly benefited from two types of study; studies of parasite–host relationships and studies of the evolution of cooperation, which, as we will see, share considerable conceptual overlap.

# 10.1 The evolution of parasite virulence

Pathogens and parasites by definition exert negative effects on the fitness of their hosts. However, they vary in the severity of those effects, such that some diseases allow their hosts to live a long time and suffer few adverse effects, while other hosts are almost instantly overcome and die. Why this variety? Earlier, conventional wisdom led many biologists to believe that parasites would eventually evolve so as not to harm their hosts, on which they depend for their persistence. But in fact the truth, as in the case of HIV (Chapter 8) is often far from this: some parasites evolve to become more virulent over time. So the story must be more complex than the conventional wisdom would suggest.

The major breakthrough in understanding was made by Anderson and May (1982). They derived a general relationship for the fitness of a **horizontally transmitted** pathogen whose population is at equilibrium with that of its host. In such cases, lifetime reproductive success, $R_0$, is an appropriate measure of fitness:

$$R_0 = \frac{\beta(N)}{\mu + \alpha + \nu}$$

where $\beta$ is the transmission rate of the disease, dependent on host density $N$, $\alpha$ is the parasite induced mortality rate of the host (virulence), $\nu$ is the host recovery rate, and $\mu$ is the death rate of uninfected hosts.

If all these parameters are independent of each other, it is clear that virulence should be zero, so as to make $R_0$ as large as possible. This is the sense behind the intuitive, but discredited, conventional wisdom. However, this assumes an absence of genetic correlations between traits. In particular, virulence and the transmission rate of the disease might be positively related. One likely source of such a correlation might be that transmission relies on

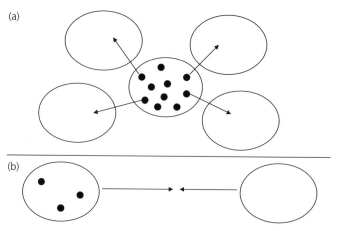

**Fig. 10.1**    Virulence and host density. High virulence (black circles) kills off hosts (large circles) rapidly, but that does not matter when hosts are common and the pathogen is easily transmitted (a). However, when hosts are rare, pathogens must be less virulent (fewer black circles) to keep their host alive long enough to encounter susceptible hosts (b).

rapid multiplication of the disease organism within the host to create many disease propagules, or inducing adverse symptoms in the host, such as coughing, that aid transmission. If this is the case, then the optimal virulence for the parasite will be such that the marginal gains of virulence, in terms of added propagule production and transmission, balance the marginal costs, in terms of reduced host survival (Figure 10.1).

The best example of evolution of virulence so as to balance these costs and benefits is that of the rabbit myxoma virus. This virus is endemic to South America but was released into Australia in 1950 to control rabbits. The rabbits had been introduced to Australia by Europeans but had multiplied to plague proportions in the absence of natural predators and disease. On release the myxoma virus did all that had been hoped; it was 99% fatal and reduced rabbit numbers to one-sixth their previous levels in just two years. After just a few years, the effectiveness of the disease had fallen, causing between 70 and 95% fatalities in infected hosts, with a reduction in the effectiveness of control (Fenner and Ratcliffe 1965). It is known that this involved evolution of the virus and not just host evolution or an acquired immunity on the part of the rabbits, because virulence was assessed by exposing the disease to control rabbits from populations that had never encountered the virus.

It is likely that the reduction in virulence seen in the myxoma virus in Australia actually had beneficial effects on transmission rates of the disease. Infected rabbits become weak and immobile and develop skin lesions that aid transmission of the virus to other individuals through mosquito vectors

(Fenner and Ratcliffe 1965). Furthermore, the mosquito vectors are rather seasonal and thus infected hosts must remain alive for extensive periods for the disease to persist. Living infected individuals are thus efficient transmitters of the virus. Although the virus may gain in the short term through more rapid overwhelming of the host, a dead host is no use as a source of transmission.

Hence, selection between hosts infected with strains of the diseases that vary in virulence can cause virulence to evolve to varying levels depending on the strength of the correlations between traits, especially between virulence and transmission rate. There is, however, another potential source of selection on virulence; selection between strains of the disease within a single host. There are two ways in which an individual can acquire more than one strain of a disease. First, an individual can acquire the disease more than once from different sources. An alternative is if the disease mutates and diversifies within the host following a single infection event. Multiple infections can then select for increased virulence. The reason is that now there is competition between unrelated individuals for a single resource. A disease strain that is prudent and is relatively benign to its host will be out-competed by a disease strain that is less prudent but exploits the resource before others can profit from it.

This tendency for a resource shared by unrelated individuals to be overexploited is known as the 'tragedy of the commons' (Hardin 1968) and is best known in the context of human overexploitation of marine fisheries. When several individuals exploit a shared host and there is a trade-off between virulence and competitiveness it can be shown that the evolutionary stable virulence strategy is $1 - r$ (Frank 1996) where $r$ is the average coefficient of relatedness between individuals (Chapter 7). Multiple infections will tend to decrease $r$ and thus increase virulence. HIV infections consist of multiple strains, and should select for increased virulence during the course of an infection and may be the ultimate cause of AIDS onset (Bull 1994).

Therefore, trade-offs between virulence and transmission rates, combined with selection between hosts, and selection within hosts between disease strains can both contribute towards the evolution of virulence. A third major variable is known to affect virulence: vertical as opposed to horizontal transmission. If pathogens are vertically transmitted, they are passed from parents to their offspring down the generations. If pathogens are horizontally transmitted they are able to pass between unrelated individuals within a single generation (Figure 10.2). If the proportion of new infections through vertical as opposed to horizontal transmission is high, it pays the pathogen to allow the host to reproduce as much as possible, so that the disease has the opportunity to infect a larger number of host offspring. In this sense the interests of the pathogen and those of the host become more closely aligned

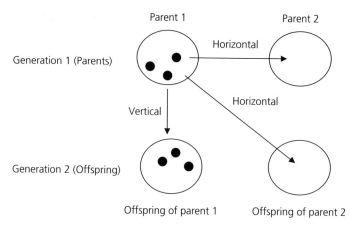

**Fig. 10.2**    Horizontal and vertical transmission of parasites (black circles) between hosts (large circles).

than through horizontal transmission. Frank (1996) has also showed that a further selective pressure may help reduce the virulence of vertically transmitted pathogen: as transmission becomes more and more exclusively vertical, the chances of multiple infections become very low. Thus, selection for virulence within hosts becomes weaker and weaker.

Is there evidence that vertical transmission selects for benevolence towards hosts? The answer is yes. In a study on fig wasps and their nematode parasites, Herre (1993) found that variation in the opportunity for horizontal as opposed to vertical transmission across species was correlated with the virulence of the nematodes. The wasp species develop in figs that the females must enter in order to lay their eggs. Wasp species differ in the typical number of females that will enter a given fig: sometimes only a single female will enter a fig (the single foundress situation), but at other times many females will enter a single fig (multiple foundresses, see also Chapter 5). After laying her eggs, the female wasp dies inside the fig and her nematodes are released to infect other fig wasps. In the single foundress situation, the nematode parasites of the fig wasp are exclusively transmitted from mother to offspring since there is only one adult wasp per fig. However, if there are multiple foundresses, the nematode progeny can infect the offspring not just of their former host, but those of other wasps, a situation more akin to horizontal transmission. Herre measured the virulence of the nematodes on the wasps by comparing the fecundity of wasps in fig fruits infected with nematodes or free of them. Across 11 species of wasp, he found a positive correlation between the proportion of fruits that were multiple foundress and the virulence of the nematodes. Frank (1996) has since argued that the variation in virulence is most likely caused by changes in the frequency of multiple infection resulting from differences in the mode of transmission, rather than the mode of transmission *per se*.

So far our discussion of the evolution of antagonism has been restricted to the relative virulence of pathogens and parasites. None of the organisms thus far mentioned are known to have positive fitness effects on their hosts, but vary in the extent of the negative effects they exert. In addition, because we have been focussing on the evolution of parasitic organisms, we have also been concerned exclusively with so-called symbiotic interactions: those that concern long-term associations between the two species relative to the life of one of them. We shall now widen the discussion to include mutualistic relationships including those that do not involve any permanent lifetime association between the two partners.

## 10.2   The origins and maintenance of mutualism

Mutualisms (e.g. Figures 10.3, 10.4) are those interactions in which both partners accrue increased fitness relative to individuals that do not engage in it. How might selection favour the expression of traits that help other individuals of a different species, given a prior relationship that was more nearly neutral or even antagonistic? Frank (1997) has suggested that two processes act to help the spread of such traits: first, an initially high level of expression on first meeting in both partners. Second, spatial association between pairs of individuals with positive synergism so that benefits can be returned to the donor (see also Yamamura *et al.* 2004).

Imagine that when two species first meet, both species possess a trait that provides a small benefit to another species at some cost to itself. If these benefits are below a threshold value, selection will favour lowering expression of the trait, such that a mutualism does not arise (Frank 1997). But if the expression of the trait is above a certain threshold, the synergistic effects between species can push expression of the trait to an equilibrium at higher levels of expression. If the benefits of the traits are high relative to the costs, the threshold value is lower and the equilibrium value higher.

Association in space is the second factor that can enhance the development of mutualism. Two types of association are in fact important: associations within populations of each species, and associations between species. Associations within species are important to raise the relatedness between individuals and prevent antagonism between them, as seen earlier in the chapter. Associations between species make it easier for the mutualism threshold to be passed—once passed, continued spatial association may not be required as long as associations within species remain fairly high. The mechanisms of spatial binding between different species are various. For

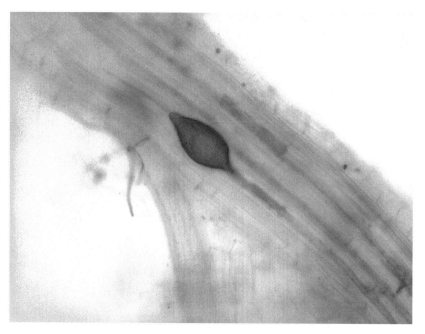

**Fig. 10.3**    A plant–mycorrhizal symbiosis. The picture shows a root of *Plantago lanceolata*, with the fungus *Glomus hoi* (stained dark) growing through it. The dark blob (50 $\mu$m across) is a vesicle, thought to be a storage organ for the fungus. The fainter rectangular structure to the right of the vesicle is an arbuscule, believed to be the site of nutrient exchange between the fungus and the plant. Photo courtesy of Angela Hodge.

example, aphids pass their mutualistic bacteria, *Buchnera* (Figure 10.4), from mother to embryo (vertical transmission). Mitochondria and chloroplasts are also transmitted vertically through gamete cytoplasm (Figure 10.2). Spatial binding of a different nature occurs between genes in a genome via linkage, and that contributes towards cooperation between genes in a cell. These latter examples illustrate that mutualism has occurred not just between what we now think of as different species, but that it was involved in the major transitions leading to modern species (Chapter 2). In many cases, the theory relevant to interspecies mutualisms has developed from thinking about these major transitions (see Maynard Smith and Szathmáry 1995; Frank 1997).

The models of Frank predict a starting threshold degree of mutual benefit below which mutualism will not evolve, and many other models make the starting assumption that mutual synergism already exists. This assumption is rather restrictive, since in many and perhaps most cases mutualism has evolved from previously antagonistic relationships. Recently, Law and Dieckmann (1998) have developed a model in which a permanent symbiotic unit with purely vertical transmission can develop from an antagonistic

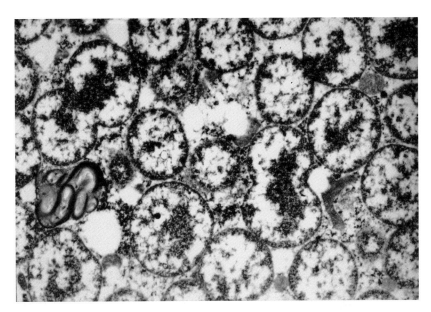

**Fig. 10.4**    Cells of the symbiotic bacterium *Buchnera*, within the cytoplasm of a type of specialized cell, bacteriocytes, within the body cavity of the pea aphid, *Acyrthosiphon pisum*. The *Buchnera* are about 4 μm across. *Buchnera* are vertically transmitted and provide essential amino acids to the aphid which are absent from their diet of plant sap. Photo courtesy of Angela Douglas.

exploiter–victim relationship (Figure 10.5). The model specifies a mortality cost to the victim in the free-living state of reducing the amount of resource donated to the exploiter in the symbiotic state. With no such cost the symbiosis breaks down: victims reduce their flow of resources to their exploiters until the species no longer interact. When the cost is sufficiently high, however, there is selection on both parties to increase the degree of coupled replication until this is the sole form of reproduction, despite the fact that one species continues to exploit the other. Such a system is one of mutual dependence, since both species now do better inside than outside the relationship, despite the fact that only one species transfers resources to the other. An experiment by Jeon (1972), involving amoebae and a parasitic bacterium, has shown the evolution of such mutual dependence in the presence of a viability cost to the amoeba in the free-living state, and despite the fact that that viability of the amoeba was no higher in symbiosis than before.

Hence, mutualism can arise in several ways from other kinds of interaction, and certain conditions make the origin of mutualism more likely. Once mutualism has arisen, however, a subsequent problem is how such mutualisms can maintain their stability over time. Where a relationship involves benefit as well as cost, the stability of the relationship is susceptible to invasion by mutant individuals that receive benefit from the other species

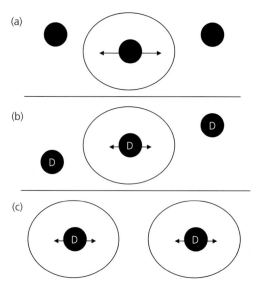

**Fig. 10.5**   The merger of lineages through exploitation. In stage (a), two species coexist, and one (large circle) occasionally exploits the other (black circle, arrows show flow of resources) in a symbiosis. However, the symbiont can also occur in the free-living state. In stage (b), the symbiont develops defenses (D) that stem the flow of resource to the host somewhat (shorter arrows). In stage (c), the free-living form of the symbiont becomes extinct because of the cost of defences in the free-living state, giving rise to a permanent exploitative symbiosis.

|  | Partner | |
|---|---|---|
|  | Cooperate | Defect |
| **Cooperate** | Reward<br><br>3 | Sucker<br><br>0 |
| **Defect** | Temptation<br><br>5 | Punishment<br><br>1 |

You

**Fig. 10.6**   The Prisoner's dilemma. Four possible interactions are shown, and the pay-off to you in arbitrary fitness units. If the partner cooperates, it is best to defect (the temptation); if the partner defects, it is best to defect (the punishment) rather than cooperate (the sucker). Result: everyone always defects, cooperation is absent.

without donating anything themselves. Theoretically, such a situation is a Prisoner's dilemma game (Figure 10.6), in which for each party it is best to defect whether the partner cooperates or defects. Thus, the Prisoner's dilemma becomes a trap in which cooperation cannot be a stable outcome. Indeed, there is considerable evidence that cheating is a common, though not exclusive outcome of mutualistic relationships. For example, many insects rob flowers of nectar without affecting pollination. The surprise is not that such cheaters exist, but that some mutualisms have survived for a long time without being destroyed by such processes—the relationship between figs and their pollinating wasps, between corals and algae, between mycorhizal fungi and terrestrial plants are all ancient. What maintains such mutualisms?

One possible answer is once again vertical transmission. Clearly, however, there are still many mutualisms that are not transmitted vertically. These include mycorrhizal associations (Figure 10.3), the rhizobia bacteria of legumes, pollination interactions (Figure 10.7), seed dispersal interactions, most gut floral associations, and cleaner associations. Many of these interactions also involve multi-species associations, which are again predicted to favour exploitation of the mutualistic partner. Clearly, there must be additional mechanisms at work that can maintain the stability of cooperation across species (Wilkinson and Sherratt 2001). There are several theoretical possibilities, but with somewhat limited supporting evidence at present.

One possible mechanism that might maintain the stability of a mutualism is if the beneficial donations are not costly to the donor. In fact, many mutualists donate products, or by-products, that are available in excess or are not limiting to them: for example, plants donate carbon, which is normally not a limiting nutrient, both to their mycorrhizal fungi and to their pollinators in the form of nectar. Lack of costs means that mutualists are playing a game more closely akin to 'nice guys win' in which it pays both partners to cooperate even in the short term because defection does not increase pay-offs.

Another potential stabilizing factor is if the partners can retaliate to defection (sanctions). Under such conditions, the decision to cooperate or defect can be made conditional on the previous actions of the other partner. It is then relatively easy to generate conditions under which persistent cooperation is the stable outcome because the lifetime pay-off, as opposed to the short-term pay-off, is now greater if organisms do not purely defect (Maynard Smith and Szathmáry 1995; Wilkinson and Sheratt 2001). One example comes from the Yucca—Yucca moth pollination mutualism (Figure 10.7). Here the pollinating moth also consumes some of the Yucca seeds. In cases where a larger number of the seeds are being consumed, some Yucca species are able to selectively abort those developing fruits (Pellmyr and Huth 1994). In this case therefore, the cost of overexploitation is paid by the moth's offspring, but that is just as effective a mechanism of selection.

**Fig. 10.7**   Two moths, *Tegeticula yuccasella*, in a yucca flower. The moth on the left is pollinating the flower, while the one on the right is laying its eggs at the base of the flower. Fruits infested with many moth larvae are more likely to be aborted by the plant, a mechanism that may help to maintain the mutualism by prevention of cheating. Photo courtesy of Olle Pellmyr.

Sanctions have recently also been demonstrated in the *Rhizobium*/legume mutualism and may help maintain it there in the absence of some other forces, such as vertical transmission (Kiers *et al.* 2003). In this case, the legume appears to withhold oxygen from root nodules containing 'cheating' bacteria that are not providing sufficient nitrogen to the plant.

Control over mixing also occurs via uniparental inheritance of organelles (Chapter 2), and via the unicellular zygote stage through which all multicellular organisms pass during their life cycle, which contributes to the stability of the body. Another form of control that selects against symbiont exploitation is reproductive fairness (Frank 1997). In meiosis and mitosis the replication of the chromosomes is rigidly controlled such that each has parity with others and an equal chance of being present in the next generation. Under such rigid control the only way that a chromosome can increase its own success is to work for the benefit of the group (the genome). In some insects that play host to symbionts, rigid control of which symbionts are transmitted to the next generation also occurs. In the sucking louse *Haematopus*, symbionts are separated early in development into a 'somatic' group which never enter the next generation and are involved in the gut symbiosis, and a 'germ' line which are destined to enter the next generation (Buchner 1965). Again, the somatic group can only increase its own success by working for the benefit of the group. A final way to enforce control is via a policing strategy that directly combats cheats. This occurs in the case of mitochondrial cheats that cause cytoplasmic male sterility in plants (Chapter 5) and whose effects are countered in some strains by nuclear genes.

Finally, while the evolution of cheating may sometimes lead to dissolution of an interaction, or extinction of species, another possible outcome is stable coexistence of cheater and mutualist. In fact stable coexistence is known from many systems, such as the yucca/yucca-moth system, and figs and fig wasps, both of which are host to non-pollinating insects. Ferriere *et al.* (2002) have used an adaptive dynamics model (Chapter 8) to investigate this possibility, assuming that symbionts compete for the resources received in the relationship in an asymmetric way such that mutant individuals are either more or less successful than residents under competition. They find that under certain conditions, evolutionary branching can occur, leading to the stable coexistence of a cheating and more mutualistic species.

Thus, a number of conditions may favour the origin and stability of mutualistic relationships, some of which are consistent with empirical data, albeit mostly anecdotal. The challenge is now to begin to test these ideas more rigidly. In addition, further theoretical and empirical research may uncover additional alternative mechanisms for maintaining the stability of cooperation between species. The next few years should be exciting in terms of furthering our understanding.

Our discussion of the evolution of species interactions in the last two chapters has largely considered only one partner in the relationship evolving at a time. Obviously though, and especially in intimate or obligate associations, each partner can serve as a selection pressure for the other. Such 'co-evolution' can then potentially lead to outcomes that differ markedly from the consideration of only a single evolving species. Co-evolution will be the subject of the next chapter.

## 10.3 Further reading

Ebert and Herre (1996) and Bronstein (2001) introduce virulence and mutualism, respectively. Frank (1996, 1997) reviews the theory in depth along with evidence.

# 11 Evolving together

In much of the last two chapters we attempted to explain the outcome of ecological interactions by assuming one species evolving in relation to constraints provided by another, evolutionarily static, species. Of course there is no reason why this process need be so one-way, and if it is not, then it is essential that the reciprocal evolution of both species in an interaction be considered. Such reciprocal evolution of interacting species is called co-evolution.

The aim of co-evolutionary studies is to predict or understand the evolutionary dynamics and evolutionary outcomes that result from species interactions. The subject actually has one of the longest histories of any in evolutionary ecology, with basic concepts and theory dating to the 1950s (see Thompson 1999). However, the intensity of research has only picked up in the last two decades, aided, particularly, by the **molecular and cladistic revolutions**. The evolutionary dynamics of interacting species do not readily lend themselves to study. Although use of the fossil record has been attempted, such studies on their own are often deficient in key data relating to the interaction. By far, the most promising progress has been made with living species, sometimes in combination with the fossil record.

To observe the dynamics and outcomes from living species, evolutionary ecologists take two approaches: a longitudinal (historical) one that reconstructs the past, normally using cross-species phylogenies to estimate the pattern of changes over time. Sometimes, however, the evolutionary dynamics are sufficiently rapid that long term field studies can detect them. A complementary (transverse) approach is to observe the interaction at different points in space, which highlights possible alternative long-term temporal outcomes. Recently, it has been suggested (Thompson 1994, 2001) that the spatial dynamics are also the key to understanding the temporal dynamics. Before we proceed to examining that hypothesis, let us have a look at some of the evidence that describes how co-evolution can proceed.

From several case studies of interacting species in the field, several alternative dynamical outcomes are now known or may be implied (Table 11.1). They are doubtless just a small sample of the co-evolutionary universe. These outcomes have previously been referred to as the 'modes of co-evolution' (Thompson 1989, 1994). I restrict myself here to modes that relate specifically

**Table 11.1** Some alternative modes of co-evolution, following Thompson (1989, 1994)

| Empirical model | Example | Properties |
|---|---|---|
| Diversifying co-evolution | Maternally inherited symbionts (Thompson 1987) | One species stimulates speciation in another |
| Escape-and-radiation co-evolution | Umbelliferae and their phytophages (Berenbaum 1983) | Radiation in one lineage occurs while it temporarily escapes the interaction with another |
| Arms race | Flower morphology and morphology of pollinators (Steiner and Whitehead 1990) | Directional change in quantitative traits, perhaps reaching stable or local equilibria |
| Co-evolutionary alternation | Cuckoos and their hosts (Davies and Brooke 1989) | The identity of interacting species changes as a result of an adaptive response in one species which initially decreases but later increases the interaction |
| Mutual dependence | Eukaryotes and their mitochondria/chloroplasts (Margulis and Bermudes 1985) | A mutualistic relationship shows temporal stability with increasing specialization and obligation |
| Co-evolutionary successional cycles | *Trifolium repens* and grasses (Turkington 1989) | Local succession creates changing patterns of association between species creating local or temporal interactions in which co-evolution occurs |
| Co-evolutionary turnover | Body size of *Anolis* lizards in the Caribbean (Roughgarden and Pacala 1989) | Asymmetric competition creates cycles of invasion, co-evolution, extinction, evolution, and reinvasion |

to outcomes (rather than assumptions). They are distinguished from each other by four characteristics (Table 11.2).

# 11.1   The number of interacting species

The first characteristic is the number of interacting species at any point in time. In some interactions this is static. For example, mutual dependence interactions (Figure 11.1) characterize mutualisms that become increasingly intimate and obligate over time. This has occurred most notably with the mitochondria and chloroplasts of eukaryotes, all three of which were

**Table 11.2** The characteristics that differentiate between the modes of co-evolution.

| Mode of co-evolution | Number of interacting species | Dynamics of species richness | Dynamics of traits | Dynamics of antagonism |
|---|---|---|---|---|
| Diversifying | 2 | Radiation during the interaction | NS | NS |
| Escape-and-radiation | (1), 2, > 2 | Radiation in absence of interaction | Punctuated equilibria | Increasing antagonism |
| Arms race | NS | NS | Dynamic, directional, local equilibria | Increasing antagonism |
| Co-evolutionary alternation | (1), 2, > 2 | Stable | Dynamic | Increasing antagonism |
| Mutual dependence | 2 | Stable | Stable | Reduced antagonism |
| Co-evolutionary successional cycles | NS | Stable | Stable | Reduced antagonism |
| Co-evolutionary turnover | (1), 2 | Cycles of invasion and extinction | Dynamic | Stable antagonism |

*Notes:* NS indicates where the characteristic is not integral to the model definition. (1) refers to the case where the interaction ceases for one or more species.

previously independent-living bacteria. Now, however, eukaryotes cannot survive without their mitochondria or chloroplasts and neither can mitochondria or chloroplasts survive independently of eukaryotic cells.

Other modes of co-evolution, however, involve a change in the number of interacting species over time. Escape-and-radiation co-evolution (Figure 11.1) refers to the process that may have occurred in many plant–herbivore interactions, whereby a plant develops a novel defence against the herbivores which ceases the interaction temporarily. In the absence of herbivory, the plants diversify into a greater number of species, perhaps because incipient species can more readily establish. Eventually, a herbivore species evolves a counter-defence, and this can then speciate into the new niches provided by the previously resistant plants. Thus, the process involves first a decrease in the number of species, and later a compensating increase. The best evidence for this process comes from the Umbelliferae family and their lepidopteran herbivores (Berenbaum 1983) (Figure 11.3). A similar dynamic occurs in co-evolutionary alternation (Figure 11.1). This occurs when a host evolves defences against a parasite, and in response the parasite is selected to look for alternative unresistant hosts. Having

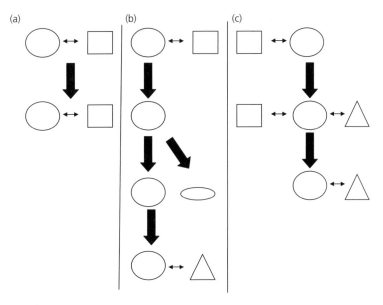

**Fig. 11.1** Co-evolution and the number of interacting species (represented by the different shapes connected by double headed arrows). In mutual dependence, this remains steady (a), In escape-and-radiation co-evolution (b), one of the interacting species goes, extinct, the other radiates and finally becomes host to another species. In co-evolutionary alternation (c), one species is added to the interaction as another is removed.

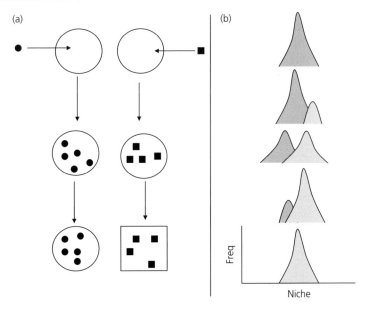

**Fig. 11.2** Speciation and extinction resulting from co-evolution. In diversifying co-evolution (a), such as between a host and a symbiont, interactions with alternative symbionts can lead to differentiation of the host. In co-evolutionary turnover (b) one species is driven extinct during the interaction (in this case via competition), but the resulting changes make the survivor vulnerable to extinction itself (in this case by subsequent competitors).

**Fig. 11.3**   A parsnip webworm caterpillar, *Depressaria pastinacella*, feeding in the flower head of a wild parsnip plant, *Pastinaca sativa*. Parsnips and webworms represent an advanced stage in escape-and-radiation co-evolution, with very nasty plant and very specialized herbivore. The parsnip contains both linear and angular furanocoumarins, which the webworm is able to detoxify. Photo courtesy of May Berenbaum and Arthur Zangerl.

found one, it eventually switches hosts entirely. Over time the old host may eventually lose its previous resistance. This process has been suggested to occur in the European cuckoo, a brood parasite that lays its eggs into the nests of passerine bird species (Davies and Brooke 1989). Again, the number of interacting species changes for both host and parasite.

Co-evolutionary turnover (Figure 11.2) is another mode of co-evolution that involves a change in the number of interacting species. This occurs when a species invades a new habitat and finds itself superior to the native competition. The invader evolves convergently to resemble the native species, forcing divergent changes on the traits of the competitor, eventually sending it extinct. Eventually, this species may fall victim to a new invader.

The process has been suggested from observations on the *Anolis* lizards of the Caribbean islands, where body size is the trait under question (Roughgarden and Pacala 1989). Invading lizards arriving at an occupied island tend to have large bodies and outcompete the smaller competitors, evolving convergently towards a smaller body size that supports the highest carrying capacity. Eventually, the native lizard may be forced into small bodied niches that cannot support large populations and make it vulnerable to extinction. Again, the number of interacting species varies over time.

## 11.2   The dynamics of species richness

The second characteristic that distinguishes the modes of co-evolution is whether speciation and extinction result from the interaction, leading to temporal dynamics of taxa. In some interactions neither speciation nor extinction is implied as a result of the co-evolutionary process. This is the case for mutual dependence. It is also the case for arms races. Arms races occur when quantitative traits in both species show reciprocal directional evolution. A well-known example is that of floral traits and those of their pollinators. For example, *Rediviva* bees in South Africa visit *Diascia* flowers to gather oils, which they collect from the base of flower spurs by inserting their forelegs (Figure 11.4). Some populations of bees have evolved extraordinary long forelegs, which match the long spurs of the flowers they visit (Steiner and Whitehead 1990). It is likely that both characters have co-evolved directionally: the bee to collect oils, and the flower to force more effective pollination service from the bee. In other cases, as we have already seen, co-evolution alters species richness. Escape-and-radiation co-evolution increases it. Another interaction that involves speciation is the aptly named 'diversifying co-evolution' (Figure 11.2). This occurs in some maternally inherited symbionts, such as *Wolbachia* and their hosts: the symbiont promotes reproductive isolation, through parthenogenesis or an incompatibility mechanism. As a result new taxa of hosts are produced, and, of course, new symbiont taxa too. As well as speciation, extinction can occur, as in co-evolutionary turnover (Figure 11.2).

## 11.3   The temporal dynamics of traits

The third characteristic of the co-evolutionary process is the temporal dynamics of traits involved in the interaction. In mutual dependence one prediction is that these dynamics are reduced since change hinders efficient

**Fig. 11.4** The bee *Rediviva neliana* collecting oil from a flower of *Diascia fetcaniensis*. The forelegs of this bee vary in length from 9 to 15 mm in different populations correlating with the length of the floral spurs of local *Diascia* flowers. Photo courtesy of Kim Steiner.

interaction between the species, leading to stabilizing selection. A possible illustration is in the mutualisms that involve ancient asexuals; they include some mycorrhizal fungi, which inhabit the roots of plants, and the fungi that are the food for leaf-cutter ants (Chapter 2). Another type of interaction that can stabilize traits involves competitive interactions between plant species that occur in co-evolutionary successional cycles. In these, succession creates patterns of association between species in which co-evolution occurs. This is best documented from white clover in grassland ecosystems. The clover becomes adapted to flourish best next to whatever species of grass it has become associated with. This is a particular kind of local adaptation. As a result, competition between the clover and the grass is ameliorated (Turkington 1989). Other interactions involve different trait dynamics. Escape-and-radiation co-evolution involves periods of trait stability and instability. Co-evolutionary arms races imply trait dynamics, although local equilibria may be achieved, and dynamics are also found in co-evolutionary turnover and co-evolutionary alternation. Diversifying co-evolution implies nothing particular about these dynamics.

## 11.4 The dynamics of antagonism

The fourth and final characteristic is the dynamics of antagonism. Many cases of co-evolution involve escalating degrees of antagonism. Included here are escape-and-radiation co-evolution, arms races, and co-evolutionary alternation. Arms races are interesting because they can occur, as in pollination interactions, within a mutualism, which serves to underline how fragile such mutualisms are to exploitation. Other modes highlight decreasing degrees of antagonism: mutual dependence and co-evolutionary successional cycles. Co-evolutionary turnover, while antagonistic, does not imply any change in the degree of antagonism during the interaction. Diversifying co-evolution does not specify any particular degree of antagonism.

Hence, co-evolution may or may not result in changes in the specificity of the interaction, in the degree of antagonism, in species richness through speciation or extinction, and variable dynamics of the traits involved. These are the things we wish to understand or predict. To do so we must make some starting assumptions and formulate models with them. The models that have attempted to predict co-evolutionary outcomes mainly address three types of interaction: competition (Chapter 9), predator–prey, and host–symbiont (Chapter 10). Some have been successful in identifying conditions under which the observed modes of co-evolution occur. As we saw in the last chapter, there is some theory that can now predict the conditions for mutual dependence in a symbiosis. With regard to competition, Roughgarden (1995, p. 110) and co-workers have extensively modelled scenarios of invasion and competition with regard to the *Anolis* lizard patterns. They find that a taxon-loop involving invasion of larger bodied species, followed by displacement and extinction of the smaller resident can be predicted if competition is asymmetric and the width of the carrying capacity small. This may then explain the patterns of body size and species richness seen on certain islands. In this model the trait, body size, is assumed to be controlled by numerous genes of small effect (quantitative genetics), which is a reasonable assumption.

Other types of interaction demand alternative genetic assumptions. Models investigating plants and their pathogens, such as rust fungi, frequently assume a single major gene locus controlling the interaction in each species, so-called 'gene-for-gene' co-evolution. For example, a pathogen may be virulent or avirulent and a host resistant or not resistant. This appears to be the case in many plant/plant–pathogen interactions. Gene-for-gene co-evolution has elsewhere been described as an alternative type of co-evolution, but it refers really to an assumption rather than an outcome. However, the assumption can generate rather characteristic outcomes: stable or fluctuating genotypic polymorphisms, or fixation of one allele in both populations. Similarly, variable dynamics are implied from field studies of plant resistance across populations (Thompson and Burdon 1992). One interesting outcome occurs when the fitness of host or parasite is frequency-dependent; that is, that it is highest when rare. Under such conditions there can be cycling of host and parasite allele frequency, so-called 'Red Queen' evolution, the type of conditions that can favour variability among offspring, and hence sexual reproduction (Chapter 2) as well as polyandry (females mating with several males). There is now increasing evidence for the frequency-dependent fitness of hosts under attack from parasites, as well as for links between polyandry or recombination, and susceptibility to parasites (Lively 2001).

The theory of predator–prey co-evolution has, in contrast, evolved largely in the absence of any background empirical data (Abrams 2001). The long timescales involved in detecting both ecological and evolutionary dynamics have been the most prohibitive hurdle. Recently, however, some studies on planktonic algae and their predators have been successful in generating short-term evolution in the algae that affect the ecological dynamics of both species (Johnson and Agrawal 2003). While this is not co-evolution, the use of microcosm systems like this holds the potential for generating useful data that can test and further develop predator–prey co-evolutionary theory.

Thus there has been some useful but limited interplay between theory and data in studies of co-evolution, though one could argue that greater potential still exists. In some cases, empirical studies have provided knowledge of outcomes or assumptions that have led to the development of theory capable of explaining them. In other cases theoretical developments have predicted outcomes that demand further empirical study.

## 11.5 The geographic mosaic

One particular empirical observation has dominated much recent thinking on co-evolutionary dynamics: the observation that species interactions are spatially differentiated between subpopulations. For example, in the South African *Rediviva* bees and *Diascia* flowers (Figure 11.4), across populations of bees there is variation in the length of the bee legs that matches the length of the local floral spurs. Thus, traits have become geographically differentiated across space. This is what John Thompson (1994, 2001) has called a selection mosaic. There may also exist across space, areas where co-evolution is more intense than in other areas, perhaps simply because of variation in

**Fig. 11.5**   A crossbill, *Loxia curvirostra*, from South Hills, Idaho, south of the Rockies, feeding on the cone of a lodgepole pine, *Pinus contorta*. In this area, devoid of red squirrels, lodgepole pine cones differ from those found in the Rockies, where red squirrels are major seed predators. The cones from South Hills are more cylindrical, and have thicker scales as a defence against crossbills. The South Hills crossbills have large, stout bills as a counter-adaptation. Photo courtesy of Craig Benkman.

the degree of overlap of geographic range. An example of such hot spots comes from a study by Benkman (1999) on lodgepole pine and crossbills in the Rocky Mountains (Figure 11.5). In areas where red squirrels occur, the pines have developed rounded cones that deter predation of seeds by squirrels. In areas lacking red squirrels, the pines have developed longer cones that are an effective defence against predation by crossbills. In these areas, the crossbills have developed larger stouter bills. Thus, these areas are co-evolutionary hot spots between pines and crossbills.

Subpopulations with divergent traits may also differ in the degree of population mixing. The crossbill populations have different songs which are likely to increase reproductive isolation. Intermediate bill morphologies caused by hybridization should also be selected against, because each population lies at its own adaptive peak (Benkman 2003). In the field, the birds from different populations mate assortatively and are beginning to differentiate genetically (Benkman 2003). Over time this may lead to speciation. These results are intuitively pleasing because crossbills seem to have differentiated across much of the northern hemisphere into a number of incipient species, and locally adaptive diversifying co-evolution provides a potentially widespread mechanism. It is additionally exciting that without considering spatial structure it would be more difficult to imagine that particular mode of co-evolution occurring.

Observations such as these led Thompson to suggest in his 'geographic mosaic theory of coevolution' that geographic structure is influential in the evolution of species interactions: 'Any theory of coevolutionary dynamics must therefore take into account this geographic structuring of most taxa and interactions' (Thompson 2001, p. 332). This stimulated many workers to examine whether linking subpopulations together in a spatial structure can lead to different predictions about co-evolution in those subpopulations than would be predicted if one just considered a single subpopulation in isolation.

So far, several studies suggest that geographic structuring can be influential at that local population scale. For example, Hochberg *et al.* (2000) developed a model of an obligate symbiont and its host, and investigated how productivity differences across landscapes influence the evolution of virulence in the symbiont. When productivity is high, the host population is high in the absence of the symbiont (a source population) and when productivity is low the host population is low (a sink population). When virulent strains of the symbiont have a competitive advantage over avirulent strains (perhaps via increased transmission), the virulent strain is most likely to be found in the source population, and the avirulent strain persists in marginal habitats that do not favour cheaters or exploiters so much. Rather interestingly, if parasites are a strong selective pressure on the evolution of sex, as suggested by Hamilton, then asexual forms should be relatively favoured in marginal habitats where virulent symbionts cannot persist. This matches empirical observations on many taxa that asexual forms are commonest in marginal habitats (Chapter 2).

It is pleasing when the addition of a single assumption, such as spatial structure, can serve to alter local predictions in the direction indicated by data. However, there is a potentially much greater prize at stake; if the addition of spatial structure can also change the global direction or mode of co-evolution over all subpopulations combined. So far, work on the geographic mosaic has not addressed this in earnest, but observations, such as seen in Rocky Mountain crossbills, suggests that it might do so, at least sometimes. The case of local co-evolution of CMS mutants and nuclear genes in *Plantago lanceolata*, leading to reproductive isolation between subpopulations (Chapter 5), provides another potential example. It is also interesting that the general theory of the evolution of mutualisms increasingly involves consideration of spatial structure (Chapter 10). Perhaps this represents some convergence of the theory of species interactions from two different perspectives.

Throughout the last three chapters it has been clear that the evolution of species interactions can result in speciation. This is a logically consistent concept, for as the next chapter will show, speciation requires rather specific ecological and environmental conditions, and other species comprise an important part of the ecological environment. The next chapter looks at the mechanisms of speciation more generally.

## 11.6   Further reading

Thompson (1989) reviews the modes of co-evolution, and Thompson (1994) updates this. Thompson (1999) reviews sources of data. Thompson (2001) reviews some recent work on the geographic mosaic. Abrams (2001) and Lively (2001) deal with predator–prey theory and parasite–host theory. A special issue of the *American Naturalist* was devoted to co-evolution in May 1999 (vol. 153 Supplement). It contains a number of reviews and case studies of interest.

# 12 Birth of species

Sympatric speciation is like the measles; everyone gets it, and we all get over it.

attributed to Theodosius Dobzhansky

Until now we have been concerned almost exclusively with evolutionary change within lineages—anagenesis. Anagenesis creates differences between lineages and hence the diversity of form and function that we observe as a major feature of our planet. It is, however, just one of the two major evolutionary processes, the other being cladogenesis. Cladogenesis is the creation of new lineages, speciation. Cladogenesis and anagenesis are jointly responsible for life's diversity: without cladogenesis, anagenesis could not realize life's potential for phenotypic diversity. Without anagenesis, as we shall see, speciation could not occur.

Biologists have identified four distinct phylogenetic processes that have created new lineages in the history of life (Figure 12.1), of which we have already discussed two. The first is the assembly of biological entities from non-biological chemical processes (Chapters 2 and 3). We have no evidence that this has occurred more than once. The second process is the fusion of two lineages into a single biological entity via symbiosis (Chapters 2, 10, and 11). This has occurred a limited number of times but has often had important consequences. The third process is hybridization: two individuals

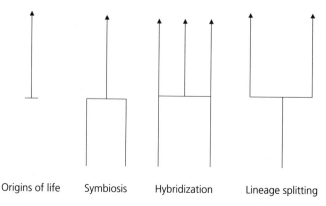

Origins of life    Symbiosis    Hybridization    Lineage splitting

**Fig. 12.1**    Four phylogenetic processes giving rise to new lineages.

from different species produce viable and fertile offspring that become reproductively isolated from their progenitors. The fourth process is when one species divides into two or more. It is these last two processes that will form the subject of this chapter.

In general biologists studying speciation are interested in three types of change: reproductive isolation, genetic differentiation, and ecological differentiation. Sometimes these act in concert, sometimes one results automatically in another, and sometimes one or more are considerably delayed. For example, in Chapter 1 we considered the case of Lake Victoria haplochromine cichlids, where reproductive isolation occurred first via changes in mate choice and coloration, followed by ecological change, while genetic differentiation leading to irreversible separation of lineages has yet to occur in many lineages. In the case of hybridization, all three processes (reproductive isolation, genetic differentiation, and ecological differentiation) may be a simultaneous consequence of the formation of hybrids.

The architects of the great Neo-Darwinian Synthesis (such as Dobzhansky, Fisher, Mayr, Simpson, Wright) laid down the groundplan for thinking about speciation. Mayr, in particular, was of the view that speciation of the fourth kind (a lineage splitting into two or more lineages) occurred allopatrically in so-called peripheral isolates, via the founder-effect, a largely non-adaptive process in which small initial isolates created a novel starting gene pool for an incipient species. Genetic differentiation could then be enhanced by divergent natural selection. He sternly denounced the possibility of sympatric speciation. Wright also advocated a large role for non-adaptive processes in his 'shifting balance' theory, whereby a small population could traverse a non-adaptive valley by **genetic drift**, and then diverge through natural selection up a different adaptive peak. Adaptive processes were simply not seen as necessary by most for speciation in allopatry, sympatric speciation was deemed unlikely at best, and natural selection could finish what other processes had started.

Since then we have gradually entered a phase in which the conceptual barriers to sympatric speciation have been eroded, the role of adaptive evolutionary change in the speciation process has become more widely appreciated, and sexual selection increasingly plays a role in our views of speciation. The relevance of speciation by hybridization has had a similarly chequered history, being more popular today than formerly. We should not fall into the trap of believing that the current vogue of research into these areas represents in some sense a greater truth. When a field swings from one extreme to another, and has yet to settle down, predicting where the equilibrium will be found is difficult and risky. There is, however, a sensible prevailing tendency to consider alternative processes more widely, and to ask not if, but in what circumstances, a process may be relevant. Data are still relatively thin on this latter point, but not totally absent, and there are good prospects for quantitative answers in the near future.

## 12.1 Speciation by hybridization

There is undeniable evidence that hybrid species are sometimes established, and indeed it has been estimated that 11% of plant species have evolved in this way. Some of the most elegant recent studies have been conducted by Rieseberg and co-workers on North American sunflowers. Two species of sunflower, *Helianthus annuus* and *Helianthus petiolaris*, have overlapping geographic ranges in western USA and have given rise to three hybrid species, *Helianthus anomalus*, *Helianthus paradoxus*, and *Helianthus deserticola* (Figure 12.2). Evidence supports a 'recombinatorial' model for the speciation of these hybrids (Figure 12.3). Under this model crossing events between the two parent species lead to novel chromosomal arrangements, as a result of chromosomal breakages (Rieseberg *et al.* 1995). The novel rearrangements cause reproductive isolation between the hybrids (Rieseberg *et al.* 1999). Some novel rearrangements carry a selective advantage, which allows invasion of new habitats (i.e. heterozygote advantage, Rieseberg *et al.* 2003). Thus, while the parent species prefer heavy clay and dry sandy soils, respectively, *H. anomalus* and *H. deserticola* are found in more xeric habitats than their parents, while *H. paradoxus* is found only in saline habitats. In the recombinatorial model then, hybridization both creates novel adaptive variation and simultaneously a mechanism for reproductive isolation.

Another route for hybridization to lead to speciation is through allopolyploidy; the creation of hybrids with double (or more) of the normal chromosome complement (Figure 12.3). Doubling of chromosome number allows a hybrid to escape infertility problems at meiosis by giving each parental chromosome its own complementary partner. The origin of several allopolyploid species has been documented in the last several decades, often as a result of species being introduced outside of their native range, leading to novel hybridizations. For example, the diploid species *Tragopogon pratensis*, *Tragopogon dubius*, and *Tragopogon porrifolius* were introduced into North America at the beginning of the twentieth century, and in about 1940 gave rise to two tetraploid hybrids in eastern Washington: *Tragopogon miscellus* (from *Tragopogon pratensis* × *Tragopogon dubius*) and *Tragopogon mirus* (from *T. dubius* × *T. porrifolius*). Both these polyploids have since expanded their range considerably (Levin 2000).

In recent years a major concern about the likelihood of hybrid speciation—the low fitness of hybrids and their consequent inability to persist—has been somewhat quelled by new data (Arnold and Emms 1998). One particularly graphic study by Peter and Rosemary Grant on Darwin's finches on the Galapagos island of Daphne Major highlights how hybrids might not always be less fit than their parent species. They studied the fitness of hybrids

**Fig. 12.2**  The sunflowers *H. annuus* and *H. petiolaris* (top), which have given rise to three hybrid species (bottom). Of these, *H. anomalus* and *H. deserticola* are found in xeric habitats, while *H. paradoxus* is found in saline habitats. Photos courtesy of Loren Rieseberg.

between three finch species (*Geospiza fuliginosa*, *Geospiza fortis*, and *Geospiza scandens*) over a number of years. In most years the hybrids were less fit than their parent species because they had intermediate beak sizes that were not best adapted to the types of seeds available (Figure 12.4). However,

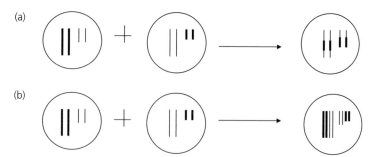

**Fig. 12.3**   Two ways in which hybridization can give rise to new species. (a) The recombinatorial model, in which chromosomal breakages occur giving rise to hybrids that are reproductively isolated from their parents. (b) Allopolyploidy, in which diploid gametes give rise to fertile hybrid offspring because each chromosome retains its complementary pair.

**Fig. 12.4**   Hybridization between Darwin's finches on the Galapagos islands. The Cactus finch, *G. scandens* (a), the medium ground finch, *G. fortis* (c), and a hybrid with an intermediate beak shape (b). In dry years, the hybrids do not survive well because their beaks are not well adapted to utilize existing seeds. In rarer wet years, when seeds are more abundant, competition is reduced and hybrids survive better. Photos courtesy of Peter Grant.

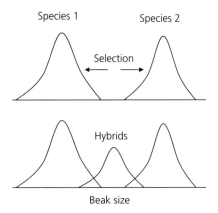

**Fig. 12.5**    Hybrids are often selected against through disruptive selection (top), and this prevents introgression of the two parent species. Occasionally, however, as in Darwin's finches, hybrids may survive due to environmental change relaxing selection against them (bottom).

in the El Niño year of 1982–83, the rainfall on the islands dramatically increased, and a dramatic change occurred in the types of seed available to the finches. In that year, the hybrids had equal or higher survivorship, recruitment, and breeding success than their parents (Grant and Grant 1993). Thus, changes in the external environment can sometimes affect drastically the fitness of parents and hybrids. It is not inconceivable that such variations in fitness might sometimes allow a hybrid species to persist and spread into new habitats (Figure 12.5).

It is also becoming clear that hybrids frequently do persist for long periods and spread despite their normally low fitness. Some populations of the closely related *Drosophila simulans* and *Drosophila mauritiana* share large segments of DNA indicating past gene exchange between them, despite experimental evidence showing strong barriers to gene exchange (Solignac and Monnerot 1986). Such observations led Arnold (1997) to suggest that the traditional assumption of low hybrid fitness, while often true, does not apply to all hybrids and that some hybrids can be fitter than their parents in a range of niches. Even if these are established rarely they can give rise to persistent lineages. Thus, the concept of hybrid speciation, not just in plants but also in animals, is experiencing something of a revival.

## 12.2    Speciation by lineage splitting

The majority of research effort on speciation has been on how lineages split into two. Speciation of this form has an important geographic context,

**Table 12.1** The geographic context of some modes of speciation (from Levin 2000)

| Size of speciating entity | Spatial relationship | | | |
|---|---|---|---|---|
| | Sympatric | Contiguous | Proximal | Distal |
| Local | Sympatric speciation | | Peripatric speciation | Disjunct Speciation |
| Big | | Parapatric speciation | | Vicariance speciation |

which has been the subject of much controversy (Table 12.1). The last two columns of Table 12.1 represent so-called allopatric speciation, speciation through geographic isolation. Allopatric speciation can be further subdivided according to the size of isolated populations and how far they are separated from each other. Under the peripatric model of peripheral isolates favoured by Mayr, the speciating entity is small, and close to the main parent population. Disjunct speciation differs in that the speciating entity is dispersed far from the original. In both these modes of speciation stochastic processes, such as founder effects and drift, may be important as the speciating population is small, though selection may also be operating to change gene frequencies. In vicariance speciation, isolating barriers emerge in an existing part of the geographic range, leading to isolation between the fragmented populations. Selection is likely to predominate here as the force causing genetic differentiation between species, and ultimately isolation, because the isolated populations are usually larger.

Some authors regard these modes of speciation as theoretically trivial, because we expect geographically isolated populations to differentiate over time by one process or another. In addition, there is good evidence that all the allopatric modes do exist in nature. For example, clear-cut cases of vicariance come from marine organisms separated by the Isthmus of Panama (approximately 3 Ma). A large number of closely related species are separated by this barrier from fishes and crabs to shrimps and sea urchins (Lessios 1998). Disjunct speciation has occurred in the 14 species of Darwin's finches on the Galapogos islands, all descended from vagrant individuals arriving from the American mainland about 3 Ma (Grant and Grant 1996). Several likely cases of peripatric speciation, occurring in birds on offshore archipelagos which differ from nearby mainland relatives are documented by Mayr (1940, 1963) in the Carribean and East Indies. Thus, allopatric speciation in general is highly plausible.

## 12.3   Splitting in sympatry?

If speciation by geographic isolation was never in doubt then the same cannot be said of sympatric speciation (Chapter 1), speciation in the absence of geographic isolation. Until very recently, there were no clear-cut empirical cases: Ernst Mayr in his 1963 book systematically attacked the best-known potential examples of the time, including the East African cichlids mentioned in Chapter 1, and another example discussed below, the apple maggot fly, as being just as (if not more) consistent with speciation in allopatry. The early theoretical work also suggested rather restrictive conditions for sympatric speciation (e.g. Maynard Smith 1966). In general, models found two problematic areas. First, it was difficult to explain assortative mating. Many models assumed strong **linkage**, or **pleiotropy** of the mating system to an ecological trait under selection, but in general pleiotropy is expected to be weak. Second, ecological differentiation is normally required to prevent competitive exclusion of one of the incipient species. Even in the presence of disruptive selection (intermediate phenotypes are selected against while extreme phenotypes are favoured), there is a tendency for just one extreme ecotype to evolve rather than two because a population will tend to move to one side or the other of the new fitness minimum. If linkage between the mating system and the ecological trait is weak, disruptive selection needs to be very strong to overcome recombination of the ecological traits.

Since then, further evidence about particular case studies has strengthened the case for sympatric speciation. In addition theoretical developments have shown that sympatric speciation is more feasible than previously thought. The theoretical advances have been (1) sexual selection leading to assortative mating (Chapters 1 and 7), and (2) adaptive dynamics leading to evolutionary branching and hence coexisting ecological polymorphism (Chapters 8 and 10) (van Doorn and Weissing 2001). Models have tended to emphasize solutions to one of these problems over the other, and hence can be termed sexual selection or ecological models.

Sexual selection models explicitly take into account the interaction of males and females and assume that males compete for mates while females exert mate choice. Normally, Fisherian runaway selection (Chapter 7) is assumed, such that male traits and female preference for those traits become genetically correlated through non-random mate choice. More extreme traits and preferences evolve until counteracted by natural selection. Mating strategies need to become polymorphic to generate reproductive isolation, and hence we need a source of disruptive selection. This might come from innate female tendencies to prefer particular divergent phenotypes, as may be the case in haplochromine cichlids. However, branching of male traits, and subsequently female preference, can also occur if rare phenotype males experience less competition for mates from other males (van Doorn and Weissing 2001, Chapter 1).

Ecological models emphasize disruptive selection on the ecological phenotype. This can readily occur in adaptive dynamic models through evolutionary branching, because the selection is not imposed externally but dependent on other resident phenotypes (frequency dependent). Under some circumstances then, evolution will drive the ecology of the species to a point where it lies at a fitness minimum and branches into a stable polymorphism (Chapter 8). Resource competition can readily do this, as long as the fitness advantage of utilizing rare resources through reduced competition outweighs the lower abundance of those resources.

So incorporation of sexual selection and ecological branching into models have shown that sympatric speciation can readily occur. In addition the evidence in favour of some of the potential examples of sexual selection has become much firmer (Chapter 1). A particularly influential case is that of the apple maggot fly *Rhagoletis pomonella* (Figure 12.6). The fly normally lays its eggs on developing apple fruit, in which the larvae develop, pupating once the fruit falls to the ground. In the Hudson valley of New York state, sometime during the mid-1800s, a host shift occurred and a new race evolved, apparently in sympatry, that developed on hawthorn fruit.

In the 1960s, Guy Bush, developed a verbal model of sympatric speciation via host race formation in the fly. Bush's model bears many resemblances to some of the classical sympatric speciation models. First, it assumes host-specific mating (a pleiotropic effect), such that each race mates and oviposits on the same fruit on which they fed as larvae. This would give assortative mating. Second, he assumed host-associated fitness trade-offs, such that a gene

(a)  (b)

**Fig. 12.6**  Apple maggot flies, *R. pomonella*, on an apple. Photos courtesy of Andrew Forbes.

giving a fitness advantage to development on one host was ill suited towards development on another host. Hybrids would be less fit than both parents on both hosts—a source of disruptive selection.

This model has subsequently been confirmed by data (Feder 1998). First, comparison of **allozymes** between the apple and hawthorn races has shown consistent differences. Races show genetic differences in host preference; females of both races prefer hawthorn but hawthorn females are more averse to apple. Once they have met, however, there is no pre-mating isolation, and no evidence of sterility or inviability barriers to reproduction. The source of the fitness trade-offs, and disruptive selection, is **phenological**: development is faster in hawthorn flies. Hawthorns fruit later in the year than apples and so hawthorn flies must develop quicker to complete development by autumn. Apple flies, however, are exposed to warmer weather before the winter arrives and must enter a deep diapause to prevent premature emergence before the winter sets in. Interestingly, when pre-winter temperatures are manipulated to expose hawthorn flies to 'apple'-like conditions, a selection response was detected in the alleles controlling diapause such that the flies became more like the apple race; in fact they did so almost completely in a single generation of selection. Thus, there is antagonistic pleiotropy between fitness determining loci, and this prevents gene flow between the races. It appears that Bush was right, and the apple maggot fly remains the single best example of sympatric speciation in action.

It remains plausible that sympatric speciation is common, but it will be frustrating to wait until a large number of organisms have been studied in the same degree of detail. Luckily, a broad-brush approach is available to determine the frequency of the different geographic modes of speciation: comparison of geographic range overlap across phylogenies (Barraclough and Vogler 2000). Under sympatric speciation, recently formed species should share large parts of their geographic range, whereas under allopatric speciation there should be almost no sharing of geographic range. Deeper down in the phylogeny, however, the degree of range overlap between sister taxa should depend on the degree of subsequent change in range change after the speciation event. By plotting the degree of overlap in the range against node height, we obtain a signal about the geographic mode of speciation (Figure 12.7).

In a comparison of several insect and vertebrate phylogenies, Barraclough and Vogler found that allopatric speciation was in fact the major signal, even in the *Rhagoletis* phylogeny where there was evidence for just a single sympatric speciation event. A further finding was that the ranges of allopatric sister taxa were frequently very different in size, suggesting that peripatric speciation is common. Broad-brush approaches like this lead us to hope that we may in a short time be able to assess the frequency of different speciation modes in a large number of taxa.

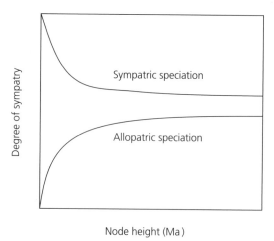

**Fig. 12.7** The degree of range overlap in species on a phylogeny, when they have been formed by either sympatric and allopatric speciation. Young species formed sympatrically have a high degree of range overlap but the overlap decreases with age due to range changes. Young allopatrically formed species show zero range overlap which increases with time due to range changes. After Barraclough *et al.* (1999).

## 12.4   Adaptive or non-adaptive differentiation?

Despite favouring different geographic modes of speciation, biologists have differed considerably over the likely processes causing the initial genetic differentiation between neospecies. Recently, there has been a dogmatic shift away from non-adaptive processes towards adaptive selection as the major process. The shift is a result of both evidence against non-adaptive processes, such as founder effects, and evidence in favour of selection.

Evidence that counters the likelihood of a role for non-adaptive processes comes from both theory and empirical evidence. For example, Barton (1998) has argued from a population genetic viewpoint that population **bottlenecks,** such that make founder effects and drift likely, are also likely to decrease the chances of incipient species persistence due to lack of genetic variation. Theoretical examination of Wright's shifting balance has shown that it is possible to cross a non-adaptive valley only if the valley is shallow and population size is small. However, very shallow non-adaptive valleys are unlikely to persist in a variable environment, so there are just as likely to be episodes when selection can drive such a peak shift. Recent experiments on *Drosophila* have also failed to produce significant population differentiation from founder events (Rundle *et al.* 1998).

The pro-selection evidence comes from several sources (Schluter 1998). First, in some adaptive radiations, ecological divergence apparently occurs as rapidly as reproductive isolation. In many freshwater fish colonizing post-glacial lakes and rivers in the Northern Hemisphere, new sympatric sister species are strongly divergent ecologically, often with one benthic and one planktivorous species coexisting (Figure 9.5). There is strong evidence that these differences are adaptive, for they correlate with increased foraging efficiency which gives an advantage in competition with other fish. In addition, sometimes mate discrimination is the result of natural selection: in sticklebacks, body size is used as a trait in mate choice, but is also one of the major traits responsible for ecological divergence. In some cases, divergent selection can lead directly to post-zygotic isolation. Some monkeyflower (*Mimulus guttatus*) populations have adapted to soils contaminated with copper (Figure 7.5), but these alleles are lethal when combined with other alleles in hybrids with other populations that are not tolerant to copper (Chapter 7).

A second area of evidence comes from examining hybridization events. If differences between species are adaptive, hybrids should be less fit than their parents in the native habitat, but this difference should disappear in the lab. Genetic mechanisms reducing hybrid fitness, however, should be apparent even in the lab. In fact, many groups produce highly viable and fertile laboratory hybrids, including *Drosophila*, Darwin's finches, East African cichlids, Hawaiian silverswords, and fishes from post-glacial lakes mentioned above. A final source of evidence comes from measuring the rate of evolution across populations or species as compared to the null hypothesis of **neutral evolution** through mutation and drift. Studies by Orr (1998) showed that of eight loci affecting male genital structure in the sister species *D. simulans* and *D. mauritiana*, all eight indicated evolution by selection. Studies on the genetic variation in quantitative traits among populations also indicate divergent selection in many species (Schluter 2000).

Thus, there is now growing evidence that the origin of species is indeed by means of natural selection, as Darwin suggested in the title of his famous book (Darwin 1859). Furthermore, lineage splitting in sympatry is looking more likely in comparison to the dogma of previous decades, and sexual selection is increasingly invoked to play a role in species divergence.

What then, is the role of ecology in the formation of species? First, ecology sets the geographic stage for speciation of whatever kind. In sympatry, speciation requires disruptive selection, which must have an ecological dimension. For hybrid speciation geographic proximity is required, and hybrid fitness and spread is determined by the ecological situation. Even in allopatric cases, selection frequently creates the initial divergences between incipient species that can lead to reproductive isolation. If stochastic processes play a role, they too are dependent on specific ecological circumstances. Finally, natural selection is a major process driving further differentiation between species such that it becomes irreversible. Thus, the birth of species has a very fundamental ecological context. In the next chapter, we will see that the death of species, extinction, primarily a population ecology phenomenon, has an evolutionary context.

## 12.5   Further reading

Schilthuizen (2001) provides a readable introduction to speciation mechanisms. Howard and Berlocher (1998) has many good chapters, such as those by Schluter (ecology), Feder (apple maggot fly), Arnold and Ems (hybrids), and Rice (genomic conflict). A special issue of *Trends in Ecology and Evolution*: July 2001. (vol. 16, no. 7), pp. 325–413 has many useful articles covering chromosomes and speciation (Riesenberg), ecology (Schluter), sexual selection (Panhuis *et al.*), and sympatry (Via). Levin (2000) covers plants well. Gavrilets (2003) reviews theory for the confident.

# 13 Death of species

It is hard to have patience with people who say 'There is no death' or 'Death doesn't matter.' There is death. And whatever is matters. And whatever happens has consequences, and it and they are irrevocable and irreversible.

C. S. Lewis

In the last chapter we saw how new evolutionary lineages may arise, increasing the number of branches on the tree of life. We will now consider how the number of branches might be depleted, thus 'pruning' the tree of life; the process known as extinction. At its limit, extinction is a population level phenomenon: it is the specific case of population dynamics when the number of individuals becomes zero. This sounds like a purely ecological process with little evolutionary perspective. Yet, as we shall see in both this and the next chapter, extinction has evolutionary causes and consequences. First, as populations decline, changes in genetic composition often occur, some with impacts on the phenotype of the organism. These changes may affect the probability of extinction, and are the subject of this chapter. Second, certain characteristics may incidentally make some species more likely to decline than others, or more vulnerable to extinction once in decline. These differences have come about through evolution. Finally, once a lineage has gone extinct, that may alter the likelihood of extinction or speciation in other species. The latter topics are dealt with in the next chapter.

There are several ways in which the processes of extinction might be classified. From a phylogenetic perspective, extinction can occur by a lineage simply ceasing to exist altogether, or alternatively by merger with another lineage (e.g. hybridization). Another framework is to consider what extrinsic processes in the environment might be causal agents (e.g. Purvis *et al.* 2000). We will explore all these aspects of extinction. Another useful and frequently-made distinction is between factors that make a species rare and factors that cause extinction once a species is rare (Soulé 1987; Caughley 1994; Lawton 1995). This will form the framework for the chapter.

## 13.1   Reasons for rarity

Rarity is most commonly defined as some combination of geographic range (small) and abundance (low) that might make a species vulnerable to extinction (see Gaston and Kunin 1997). Just because a species is rare, however, does not mean it is declining. Species may be rare at three stages in their evolution: first, they may just have speciated and be in the process of expansion. Second, they may have expanded their geographic range as far as possible, but still remain rare. Third, they may be rare because their range has declined from its maximum extent (Figure 13.1).

The danger of extinction while a species is still young is illustrated by the fates of hybrid plant species that have formed in historical times (Chapter 12). The Malheur Wire Lettuce, *Stephanomeria malheurensis*, is a recent derivative of *Stephanomeria exigua* known only from a single hilltop in Oregon (Figure 13.2). The species is probably now extinct, largely due to changes in the habitat and invasion of an introduced grass, *Bromus tectorum*.

A contrasting story is that of the cordgrass *Spartina anglica*, a tetraploid formed through hybridization of *Spartina maritima* and *Spartina alternifolia*, and first recorded in Lymington, UK, in 1892. Both the hybrid and its parents grow in coastal mudflats. *S. anglica* spread rapidly following its discovery, and in a few decades had spread along much of the southern English coast and onto the French coast. It has since been introduced to other European countries as well as China and Australia to stabilize low lying coastal zones. The species cannot be considered at risk of extinction, and one factor favouring its long-term survival has been its rapid exit from the vulnerable early stage of its formation.

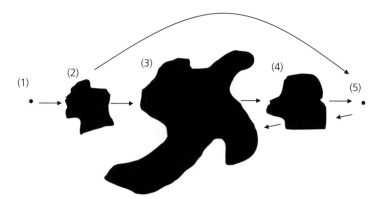

**Fig. 13.1**   Changes in geographic range during a species' lifespan, giving rise to rarity. Species are generally rare at their origination (1) and at the end of their lifespan (5), but in between may be either widespread (3) or local (2), (4).

**Fig. 13.2**   Malheur Wire Lettuce, *S. malheurensis*, at its only known site in Oregon. Photo courtesy of
Leslie Gottlieb.

Therefore, species that survive their initial rarity, like *S. anglica*, may enter
a second stage in their evolution, during which they expand their range to
some maximum because adaptations allow them to sustain positive rates of
increase beyond the area of their birth. Dispersal is obviously required
(Chapter 6) as are broad ecological tolerances (Chapter 8). Some species,
so-called endemics, will, because of their site of origin or specialization, never
attain a large geographic range. At least 44% of vascular plant species and
35% of vertebrates are endemic to 25 biodiversity 'hot spots' which represent
only 1.4% of the land area of the globe (Myers *et al.* 2000). Many of these hot
spots include island archipelagos, such as the Pacific islands, the Carribean,
New Zealand, the East Indies, and the Indian Ocean islands. Evolutionary

changes may occur among rare species that make range expansion less likely. For example, island species may evolve reductions in dispersal ability (Roff 1994; Grant 1998), such as flightlessness in birds (see Figure 13.3) and insects, and reduction in seed dispersal in plants (Chapter 6), that reduce their chances of increasing their geographic range. Rare species may also develop adaptations that favour persistence. Kunin and Schmida (1997) have identified changes to plant breeding systems and flower architecture as possible adaptations to rarity. They tested the breeding systems of 52 Israeli crucifers against their abundance and found that species with sparse populations tended to be more self-fertile. In addition, rare self-incompatible plants had unusually large flowers, which should enhance the chances of cross-pollination.

After expanding their range to some maximum, species might also become rare later in their evolutionary history through declining abundance or geographic range. The agents of decline can be either biotic or abiotic. Biotic agents of decline are extremely well illustrated by the recent effects of introduced species. The Lord Howe woodhen, *Tricholimnas sylvestris* (Figure 13.3), for example, is a flightless rail that lives on Lord Howe Island, about 600 km East of the Australian mainland. The island was first discovered in 1788 and settled in 1834, with the consequent introduction of

**Fig. 13.3**  Two Lord Howe Woodhens, *T. sylvestris*, a species of flightless rail restricted to Lord Howe Island in the South Pacific. Photo courtesy of Ian Hutton.

pigs, dogs, cats, goats, and black rats. By 1853, the woodhen was restricted to mountainous parts of the islands, and by 1920 the population was almost entirely restricted to one single mountain top of 25 ha, containing no more than 10 breeding territories. Miller found that the range of feral pigs did not overlap with the range of the woodhen and could have been the cause of decline through predation on nesting adults and eggs (Miller and Mullette 1985). After removal of the pigs, and a subsequent release of captive bred birds into other areas of the island soon filled the entire available habitat on the island. Declines due to introduced species illustrate the importance of co-evolutionary forces (Chapter 11) on species abundance. Many island birds, for example, are vulnerable to introduced ground predators because they have evolved adaptations in the absence of predators that become detrimental in presence of predators. These adaptations include flightlessness, ground-nesting, and lack of escape responses (Grant 1998).

Abiotic changes will typically involve changes in climate. Many species have experienced well-documented contractions or expansions in their range during the last several thousand years as a result of changes in global temperature associated with the glacial and interglacial periods. The redwood trees *Sequioadendron giganteum* and *Sequoia sempervirens* were widespread in North America before the Pleistocene glaciations, but the former is now restricted to a few valleys in the Sierra Nevada mountains in California, and the latter to the fog belt of the coastal ranges. *Aphodius holderi* was the most abundant dung beetle in Britain and other parts of Eurasia during the colder parts of the last glaciation, but today is restricted to a very small area in the high plateau of Tibet (Coope 1973).

Thus, at any stage of their evolutionary lifespan species may become rare. Most will be rare at their origins, but if they can expand rapidly their long-term prospects are good. Many species, however, may not ever achieve large geographic ranges or population sizes, a contributory factor to which will be their evolved characteristics. Finally, some species will experience declines after reaching their range or abundance maximum, and one contributory factor may be the evolutionary interactions with other species. When populations are rare, what might deliver the final *coup de grace*?

## 13.2   Extinction when rare

A variety of processes may eventually cause population extinction once the population has reached a small size or range. The importance of these processes can vary with the size of the population and the characteristics of the species. Many of these processes are stochastic, meaning they are not exactly predictable but occur with a probability that can be known. Others are more predictable or 'deterministic'. The processes can be environmental, demographic, or genetic in nature (Lande 1988). We will deal with these in turn.

Environmental stochasticity is the name given to unpredictable events that affect all individuals in the population in similar ways. In any given year, for example, weather conditions might promote or reduce reproduction; and biotic agents might favour or hinder survival. In one well-documented case, increases in environmental stochasticity have been strongly implicated in population extinction. The Bay subspecies of Edith's Checkerspot butterfly (*Euphydryas editha bayensis*) (see Figure 9.4) has recently experienced a number of local population extinctions. This has been linked with increasing variability in rainfall during recent decades. In dry years, the larvae cannot develop sufficiently fast to enter diapause in the summer before their food plants die, and there is massive larval mortality (McLaughlin *et al.* 2002).

Other stochastic events affecting whole populations have been implicated in recent extinctions: the last population of the Heath Hen (a type of grouse once endemic to North America) having been restricted by hunting and habitat destruction to the island of Martha's Vineyard at the turn of the century, experienced a succession of unlucky events, including a drought and fire, and unusually high predation by Goshawks. The population never recovered. The risk of extinction from purely environmental stochasticity can be modelled very simply by describing time to extinction versus population size (carrying capacity) when the average vital rates are drawn at random from a distribution (Lande 1993). Time to extinction depends on the intrinsic rate of increase of the population, being larger when rate of increase is large, but tends to asymptote with population size (Figure 13.4). Thus, it remains an effective cause of extinction even in moderately sized populations.

Demographic stochasticity in contrast is the result of changes in the average vital rates of a population due to differential success of individuals. Some years by chance most of the individuals will be lucky, and at other times unlucky, and these effects can be modelled as sampling variances drawn from a distribution. The variance in average rates is inversely proportional to the population size, and so time to extinction rises quickly (exponentially) with population size (Figure 13.4). As a consequence, unlike environmental stochasticity, demographic stochasticity is most important in very small populations.

Other demographic causes of extinction may be more deterministic in nature. The Allee effect refers to a deterministic decline in fitness with density due to non-genetic reasons. In a population already struggling to maintain positive rates of increase, the Allee effect can lead to an irreversible decline: below a critical density, fitness is reduced, reducing density still further,

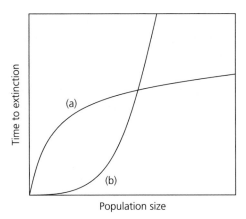

**Fig. 13.4** Time to extinction against population size. In (a), the line is asymptotic, so even quite large populations may be at risk. In (b), small populations are at risk, but even medium sized populations are relatively safe.

reduces fitness still further. Two typical biological causes are density-dependent mating success, and synergistic social interactions, such as group foraging or defence from predators. The edge effect is a related process that occurs when individuals at the edge of a population experience declining fitness. As the geographic range of a population diminishes, a greater proportion of individuals is near the edge of the range and experience this decline in fitness. In plants a typical cause of this non-genetic decline in fitness is a decline in cross-pollination. In the rare Australian plant *Banksia goodii*, for example, which grows about 20 cm high, the number of seeds set per plant is positively correlated with population size, with populations covering less than 25 m² setting no seed at all (Levin 2000). Only about 1000 plants of this species remain and most populations are very small.

A final demographic cause of extinction is a change in the balance between colonization and extinction of habitat patches or subpopulations. Networks of ephemeral populations linked by emigration and immigration, and persisting through a balance between the colonization and extinction of subpopulations, are called 'metapopulations'. Critical factors allowing the persistence of such a metapopulation include the number, size and distribution of patches, and the dispersal rates between them. As a population declines, either in the size or number of subpopulations, this can affect the balance between local extinction and local colonization. As a result, below a critical number of patches or size of subpopulations the whole system collapses inevitably towards extinction. A system where these effects have been modelled is in the Northern Spotted Owl *Strix cavrinea occidentalis* (Lande 1988). The patches here are owl territories of 2.5–8 km², which the

**Fig. 13.5**   The Glanville Fritillary butterfly, *Melitaea cinxia*. Populations on the Åland Islands in Finland are more likely to be inbred if small, and more likely to go extinct if inbred. Photo courtesy of Niclas Fritzén.

owls occupy as monogamous pairs in old growth conifer forest in the Pacific North-west. Plans for the conservation of the owl had originally envisaged conserving 500 pairs to maintain sufficient genetic variation (see below). A model based on habitat occupancy, however, showed that 500 pairs were insufficient to halt extinction from demographic causes because colonization of new territories would not balance loss of occupied territories. A similar and more recent study on separate metapopulations of Glanville Fritillary butterflies (Figure 13.5) in Finland showed that metapopulations of a few small well separated habitat patches had gone extinct during recent years while larger metapopulations with greater numbers of large proximate habitat patches persisted (Hanski and Ovaskainen 2000) as predicted by a simple model.

## 13.3   Genetic causes of extinction

Genetic factors leading to extinction include evolutionary changes in small populations. If a species with a small surviving population is in close proximity to more abundant congeners with which it can interbreed, a real

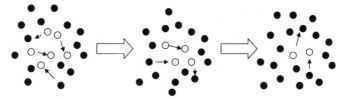

**Fig. 13.6**   Extinction via hybridization. Here a rare species (white circles) has its range surrounded by another species (black circles) with which it can hybridize, reducing its ability to replace itself. Interbreeding occurs with close-neighbours, and eventually, all the reproduction of the rare species is via hybridization.

risk of extinction comes from interspecific hybridization. There are several reasons why this may be a threat to the less abundant species (Levin 2000). First, crossing with other species reduces its potential to replenish its numbers because some reproductive potential is diverted to hybrid offspring rather than pure-bred offspring (Figure 13.6). Because the small population is less effective as a reproductive donor than the large population, the effect on its reproductive potential is much greater. Second, the hybrids might compete successfully against the parents reducing their fitness. Again, the proportional effect is greater in smaller populations. Hybridization can also increase pressure from natural enemies. Many plant hybrids are more susceptible to pest exploitation than their parents and hence can support large pest populations in proximity to their parent species (Levin 2000).

Finally, genetic introgression may be so one-way that the smaller population simply becomes absorbed into the larger one. One likely future extinction from hybridization comes from the Catalina Mahogany tree, *Cercocarpus traskiae* (Rosaceae) which occurs on Santa Catalina Island off the Californian coast. It is restricted to a single population and is hybridizing extensively with its close relative *Cercocarpus betuloides*. Eventually, this may lead to total introgression of the mahogany such that no pure-bred plants remain. There are several other botanical examples, but hybrid extinction is also a possibility in animal populations. We encountered one example from the haplochromine cichlids of Lake Victoria (Chapter 1) where mating barriers in sympatry are breaking down due to water turbidity. Another animal example comes from the endangered European White-headed duck, which has been hybridizing extensively with Ruddy ducks, introduced from North America and now considerably more abundant. Just as hybridization can create species, so can it destroy them!

Inbreeding depression is another genetic process that may cause extinction. This occurs when declining populations experience a decline in fitness through increasing homozygosity of deleterious recessive alleles. It is normally expressed through reduced fecundity and offspring viability. Inbreeding depression is readily created experimentally in captive populations. In *Drosophila melanogaster*, about half the loss in fitness is due to recessive lethal/semilethal mutations at about 5000 loci. The remainder is due to many slightly deleterious alleles that are mildly recessive (Lande 1988). When mating occurs among relatives, the chances of an offspring being homozygous for these recessive alleles are increased, and their deleterious effects are expressed. Inbreeding depression is not an automatic consequence of small population size: gradual inbreeding tends to create little depression of fitness because there is more opportunity for selection to purge the population of homozygotes, thus removing the deleterious alleles. Many insect and plant species consequently inbreed with little sign of fitness depression. In contrast, when a population suddenly declines the effects of inbreeding depression are more marked. As with Allee effects, the loss in fitness can create a positive feedback 'extinction vortex', whereby in a small population inbreeding occurs, fitness is reduced, which reduces population size, which in turn increases inbreeding which reduces fitness and so on.

Ralls *et al.* (1979) looked for signs of inbreeding depression by measuring juvenile survival in ungulate populations in captivity kept in either inbred or outbred conditions. In 41 of 44 species examined, juvenile survival was lower in the inbred populations. Inbreeding depression has also been inferred from genetic and fitness studies of wild populations. The Chihuahua spruce (Figure 13.7) was once widespread in Mexico but its range has contracted into a few isolated populations of between 15 and 2400 trees. There is very low gene flow between populations, and parents are more closely related than half-sibs would be in a typical outbred population. Forty-five per cent of seeds do not contain embryos, which is likely due to inbreeding depression (Ledig *et al.* 1997). In some cases population extinction can be attributed to inbreeding. In 42 subpopulations of the Glanville Fritillary (Figure 13.5) in Finland, the risk of extinction was negatively correlated with the proportion of heterozygosity in the populations (Saccheri *et al.* 1998), and one generation of full sib-mating in the laboratory led to substantial inbreeding depression.

In several wild populations the effects of inbreeding depression have been reversed by transfer of individuals between populations (Hedrick and Kalinowski 2000; Hedrick 2001). The Florida panther (*Felis concolor coryi*), a subspecies of mountain lion, has been isolated in southern Florida since the 1900s and the current **effective population size** in recent years has been 25 or less. The males showed a high frequency of undescended testes and deformed sperm. After the introduction of females from Texas in 1995, 14 offspring were produced from crosses with the Florida population, with a reduced frequency of sperm defects.

**Fig. 13.7** A stand of Chihuahua Spruce, *Picea chihuahuana*, at El Realito, Chihuahua, Mexico. The trees, once widespread, are now found in small isolated populations which show signs of inbreeding depression. Photo courtesy of Thomas Ledig.

How big must a population be to be safe from inbreeding depression? Franklin (1980) suggested the simple rule that an effective population size of 50 or less is at risk. This was based on very simple logic: in the absence of selection the **inbreeding coefficient** increases by $1/2Ne$ per generation due to random genetic drift, where $Ne$ is the effective population size. Therefore, to limit inbreeding to 1% per generation, a level tolerated in domesticated animals, the effective population size should be 50. It is, however, only a rough guide, and as Hedrick (2001) has pointed out, there are several cases of endangered species, such as the Californian Condor, that have had founder numbers of much less than 50 and that have rebounded from the brink of extinction without apparent inbreeding depression.

Two other genetic changes can lower the survival prospects of small populations: loss of genetic variation (and hence evolvability) through drift, and accumulation of deleterious mutations of small effect, also by drift. In small populations, random fluctuations in gene frequency (drift) tend to reduce genetic variation, leading to increased homozygosity and consequent loss of adaptability. The process countering this loss of variation is mutation, such that at any population size an equilibrium level of genetic variation is achieved, which is lower for small populations. Franklin (1980) suggested that an effective population size of 500 was sufficient to maintain genetic variation that balanced the variation in the environment that necessitates future evolutionary potential. This was based on data from *Drosophila* bristle numbers suggesting that the rate of production of genetic variation in this trait is about one thousandth of the environmental variance. However, Lande (1995) has suggested that this rule of thumb may considerably over-estimate the amount of genetic variation maintained. Recent data on *Drosophila* suggest that only about 10% of the mutational variance is in fact neutral or quasineutral (thus, replacing that lost through drift), 10 times less than assumed above. The population size required to maintain sufficient variation would then be 10 times as great (5,000).

In small populations, slightly deleterious mutations can also become fixed by chance, a source of genetic stochasticity. Individually, these mutations do not affect the risk of extinction, but collectively they can. Lande (1995) found that if no selection is assumed (just mutation and drift), mean time to extinction from such new mutations is an exponential function of population size (Figure 13.4). Thus, as for demographic stochasticity, at reasonably large population sizes ($Ne > 100$), there is little risk of extinction from genetic stochasticity. However, also assuming selection decreases the mean time to extinction because it becomes a power of (asymptotically related to) population size (Figure 13.4). As a result, genetic stochasticity might be much more important at large population sizes, comparable to environmental stochasticity, making very large populations necessary for long-term viability.

Thus, a range of processes, both evolutionary and ecological, may contribute towards extinction. These work both by making species rare and by causing extinction once rare. Evolution can determine whether a species becomes rare indirectly via selection on the ecological niche, and due to co-evolutionary interactions with other species. Among small populations, a range of evolutionary processes, such as drift, inbreeding depression, and introgression, may finally cause extinction. Many of these may only be important in very small populations, but others, such as loss of fitness due to drift, may be active in comparatively large populations. In the next chapter, we examine in more detail the characteristics of taxa that might be associated with extinction, speciation, and species richness.

## 13.4    Further reading

Caughley (1994) introduces extinction theory and explores many case studies. Lande (1988) and Lawton (1995) also cover useful concepts. Levin (2000) covers plants. Hedrick and Kalinowski (2000) and Hedrick (2001) cover inbreeding.

# 14 Big evolution

As long as you're going to be thinking anyway, think big.

Donald Trump

Up until now the book has been concerned with the properties of individual lineages and how they change through time: so-called 'microevolution'. Our focus in microevolution has been mostly on how the average properties of a species may change and why they were of a certain observed value as opposed to some other hypothetical values. This is a natural stance to take, because intraspecific variation tends to be small in comparison with interspecific variation, hence, we can justify to some extent treating species as single points in the moving cloud of phenotype space (though see Chapter 11). Evolutionary biologists can also consider the properties of clades of lineages: so-called 'macroevolution' (Figure 14.1). Macroevolutionists tend to

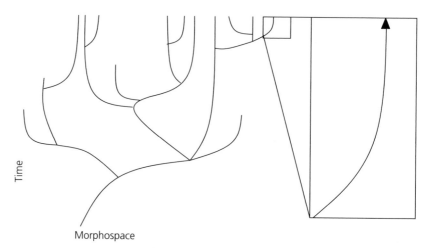

**Fig. 14.1** Macroevolution versus microevolution. The main picture depicts a clade evolving through time and morphospace. Species form and go extinct, and anagenesis occurs within species, and this clade varies through time in species richness and disparity, the two major macroevolutionary variables. Microevolution, in contrast, is concerned with individual species, and whether they speciate, go extinct, or change in form (magnified section).

concentrate not on averages but on variation and diversity. Two types of diversity, in particular, are of interest: the species richness of the clade (number of lineages) and their diversity of form (disparity). While a single species can in theory have a disparity, diversity of form among a number of species can be much greater because the absence of interbreeding among them means that evolutionary divergence can much more readily occur. Species richness is a property that is obviously only interesting in clades as opposed to individual species!

Underlying a clade's species richness are the two cladogenetic processes of speciation and extinction, the net effect of which (speciation minus extinction rate) gives rise to the net rate of diversification. Disparity is influenced not only by these cladogenetic processes, which sprouts and prunes out the different phenotypes, but also by the other evolutionary process of anagenesis, which moves the phenotypes of each lineage around. Macroevolution is of interest to ecologists simply because the currencies of interest to macroevolutionists, species richness, and diversity of form, are also currencies that ecologists measure. Thus, ecologists attempting to explain numbers of species or diversity of function frequently experience the need for macroevolutionary explanation.

We can ask a number of questions about these properties and processes affecting clades. How do they vary through time? How do they vary geographically through space? How do they vary across clades? What properties of clades and the environment affect them? To answer these questions we have the tried and tested combination of theory and data. We will discuss the contribution of theory later in the chapter. The data come from two main sources: the fossil record, and studies of extant clades. The studies of extant clades come in three main kinds. First, studies of phylogenetic tree shape enable us to derive information about cladogenetic processes. Second, studies of endangered or recently extinct species reveal underlying patterns in extinction rates. Third, studies of recent adaptive radiations allow us to infer the relative progress of diversity and disparity during a clade's evolution. It will be obvious from this discussion that not all sources of data impinge on every question. In addition, some questions have received a relative dearth of attention. In particular, most work on macroevolution has concentrated on diversity rather than disparity, as will be obvious below (Table 14.1).

## 14.1   Macroevolutionary theory

Imagine a monophyletic clade evolving. What parameters are necessary to describe its evolution? At minimum, we must describe the number of species, but we may also wish to include a description of its disparity. Since describing

**Table 14.1** Sources of macroevolutionary data and questions they can address.

| Clade property | Question | Fossil record | Tree shape | Historical extinctions | Extant radiations |
|---|---|---|---|---|---|
| Disparity | When? | ✓ | | | ✓ |
| | Where? | (✓) | | | ✓ |
| | Who? | (✓) | | | (✓) |
| Speciation | When? | ✓ | (✓) | | |
| | Where? | ✓ | (✓) | | |
| | Who? | ✓ | ✓ | | |
| Extinction | When? | ✓ | (✓) | | |
| | Where? | ✓ | (✓) | ✓ | |
| | Who? | ✓ | ✓ | ✓ | |
| Net diversification | When? | ✓ | ✓ | | ✓ |
| | Where? | ✓ | ✓ | | |
| | Who? | ✓ | ✓ | | |

*Notes:* Brackets indicate current absence of studies.

disparity requires at minimum a description of cladogenesis, we will start by simply describing the number of species and then add in disparity.

In essence the description of the species richness of a clade is very similar to describing the number of individuals in a population. A clade starts off with a single species, which speciates into two or more, which can also speciate. If this is the only process acting on the clade, the clade multiplies exponentially at a rate determined by the speciation rate. It is then simple to describe the relationship between three parameters: the species richness of the clade, its speciation rate, and the age of the clade. Knowledge of any two of these parameters allows us to calculate the third. This simple exponential growth model assumes a constant-rate deterministic speciation process with no extinction. Of course, we can assume extinction without explicitly modelling it, by just renaming the speciation parameter the net rate of diversification, equal to the speciation rate minus the extinction rate.

This model has rather uninteresting dynamics but has two useful applications. First, if a clade's growth conforms to it, it is an indication that no further complex processes are at work, and therefore that the rate of diversification is constant through time. By adding in the assumption of stochasticity in rate, we can ask if the diversities of two clades imply simply stochasticity in the same underlying rate of diversification, or if it implies different underlying rates. Thus, this simple exponential model allows us to answer two questions about the net rate of cladogenesis: is it constant (when) and does it vary between clades (who)? The model was extensively used by Stanley (1979) to compare radiation rates in different taxa, but originates much earlier in work by the mathematician Yule (1924).

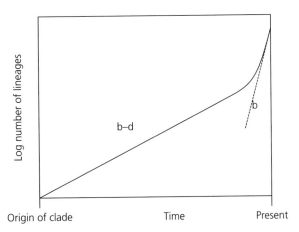

**Fig. 14.2** How to estimate speciation and extinction rates from a phylogeny of extant species. In a semi-log plot of the number of lineages over time, the log number of lineages rises at a characteristic rate, equal to the speciation minus the extinction rate (b − d). Species that are newly formed, however, have not had a chance to go extinct, so the rate of formation of species near the present is simply the speciation rate (b). The difference between the two slopes is therefore the extinction rate (d). After Harvey *et al.* (1994), with permission from the Society for the Study of Evolution.

A further simple step is to add in two discrete parameters describing the rate of speciation and extinction separately. A clade whose speciation rate is greater than its extinction rate grows in a roughly, but not exactly, exponential fashion. Over much of its history growth rate is constant. However, early on in its history its growth rate is faster, because extinction can only act on lineages that have come into existence through speciation. This early rate of increase represents the speciation rate unfettered by extinction. For clades whose past history can only be inferred from the phylogeny of extant species, this speciation rate can also be estimated as the rate of increase in branching in the phylogeny near to the present (Figure 14.2), once again these species are newly formed and have not had a chance to go extinct (Harvey *et al.* 1994). Once the speciation rate is known, the 'normal' rate of increase allows the net rate of diversification to be calculated, and the difference between the two is the extinction rate. Hence, knowledge of the dynamics of clade growth allows the rate of speciation and extinction to be calculated from the phylogenies of extant species. These rates can be calculated more simply from the fossil record because origination and extinction are more directly observable.

While a clade is radiating at constant (exponential) rate, what happens to its morphological diversity? Perhaps the simplest model assumes that anagenesis can occur in a random walk fashion and that extinction and speciation is random with respect to morphology. If morphological traits are assumed to be binary, as is practically the case with most morphological traits, then as the clade radiates exponentially, its morphological diversity increases rapidly at first but then asymptotes (Gavrilets 1999). The asymptote is caused simply by the fact that the binary nature of the traits places strong (geometric) constraints on continued morphological diversification. For continuous traits, however, which are less constrained geometrically, disparity continues to increase in a linear fashion (Foote 1996).

It seems unlikely in a finite environment that clades will grow unhindered for ever at a constant rate. The simplest model that puts some constraint on exponential growth is the so-called logistic model of Verhulst (1845), whereby as the clade approaches a theoretical carrying capacity its net rate of diversification is reduced by a feedback parameter. A wide variety of dynamics are possible in the logistic model depending on the value of the feedback parameter and the net rate of diversification, from stable equilibrium through to chaotic dynamics. In addition, interactions between taxa are possible within a logistic framework if the feedback is the result not just of the clade's own species richness but that of other clades too. This now allows displacement dynamics whereby taxa can replace each other. More complex models are possible but their necessity depends on the data to which we shall now turn.

## 14.2 Trends in time

The most extensive data on temporal trends in macroevolution are from the fossil record, in particular, that of shelly marine invertebrates, which has been extensively documented. The data tend not to be analysed at the species level because of various inaccuracies and biases that are accentuated at low taxonomic levels.

The gross history of this group (Figure 14.3) shows a generally increasing trend in family richness punctuated by sudden decreases, in particular, the five big mass extinctions occurring at the end Ordovician, late Devonian, end Permian, end Triassic, and end Cretaceous. These episodes were the end of many groups. Over time, however, virtually every possible history of family richness is demonstrated, from sudden increase and decline, to slow increase and slow decline. Decreases in diversity following the mass extinctions are then followed by increases in diversity, and this increase often takes the form of a replacement of one taxon by another, a trend that suggests that interactions between taxa are important in macroevolutionary dynamics, and further that rates of diversification are limited in a logistic sense.

Extinction rates and speciation rates both exhibit a general decline over time. This decline coincides with the replacement of taxa that exhibit

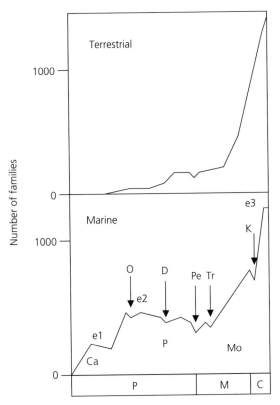

**Fig. 14.3**  How diversity has changed over time (P—Palaeozoic, M—Mesozoic, C—Cenozoic), according to the fossil record. The well-known marine family record shows three possible equilibria (e1, e2, e3) corresponding to the dominance of three faunas (Ca—Cambrian, P—Palaeozoic, and M—Modern). The equilibria are punctuated by the five big mass extinction events (O—Ordovician, D—Devonian, Pe—Permian, Tr—Triassic, and K—Cretaceous). The terrestrial record in contrast looks much more exponential. After Benton (1995) with permission from AAAS.

high rates of diversification and extinction and that are present at the start of the **Phanerozoic** (the Cambrian fauna of **trilobites** (Figure 14.4), **Monoplacophora** and **graptolites**), the emergence of a second fauna that dominates until the end of the Permian (the Palaeozoic fauna of crinoids, **cephalopods**, and **ostracods**). This fauna displays intermediate speciation and extinction rates. Finally, a modern fauna of gastropods, bivalve molluscs, and **echinoids** dominates to the present with low rates of speciation and extinction (Sepkoski 1999). A logistic model of clade growth of these three faunas, with interaction purely through the logistic feedback term, with these estimated diversification rates can predict the pattern of replacement very accurately. For best accuracy the mass extinctions need to be imposed on the system, but the pattern of faunal replacement and general trends in

**Fig. 14.4**    The trilobite *Dicranurus monstrosus*, from the Devonian of Morocco, length about 4.5 cm minus spines. Trilobites were a component of the 'Cambrian Fauna' one of the three great marine faunas in the fossil record, characterized by a high origination rate and high extinction rate. Photo courtesy of Richard Fortey and Claire Mellish, © The Natural History Museum, London.

species richness occurs even without these. In fact the net effect of the mass extinctions is to delay the replacements rather than to speed it up as is generally assumed (Kitchell and Carr 1985). Thus, the general large-scale trends for the marine realm suggest logistic growth of clades with interaction and replacement (Figure 14.3). Consistent with this, the patterns within clades show a slow decline in rates of origination over time. There is also some suggestion from the shape of phylogenies of extant taxa that rates of diversification tend to decline over time. The pattern of origination of bird families is not consistent with a constant rate model of diversification, and suggests, instead, that there has been a decline over time (Harvey *et al.* 1991). However, the terrestrial fossil family-curve looks much more exponential (Figure 14.3), and it also has been suggested that an exponential species-curve may underlie the logistic marine family-curve (Benton 1995). It does appear, therefore, that not all groups have diversified in a logistic fashion, and that different biotic realms may impose different constraints to diversification.

What of disparity? Again, variable patterns are seen in the fossil record. In some groups, such as the Palaeozoic trilobites, crinoids, and insects, diversity rose initially followed at length by a rise in disparity. In many groups however, including the Cambrian arthropods as a whole, many plant taxa, Carboniferous **ammonoids**, and Paleozoic gastropods, disparity peaks early and taxonomic richness rises afterwards (Foote 1997). This latter pattern is perhaps the most common (cf. Schluter 2000, pp. 59–60), and is *prima facie* consistent with the null pattern predicted by Gavrilets (1999). Foote (1997), however, argues that for several reasons geometric constraints are unlikely to be the only reason for the frequency of this pattern.

The relationship between disparity and taxonomic diversity can also be addressed by comparisons of recent radiations. Schluter (2000) used comparisons of replicate radiations in finches, mammals, cichlid fishes, warblers, and *Anolis* lizards to show that morphological diversity was greater in the older clades, showing a continuing rise in disparity with time. However, many of these clades do not show increased species richness with age, suggesting that taxonomic richness reaches limits first, as paralleled in some fossil groups. Studies of extant radiations also suggest that the evolutionary rates both of morphological and taxonomic richness are higher when ecological opportunity is greatest (Schluter 2000). For example, island radiations often display greater morphological diversity and species richness than their mainland counterparts.

# 14.3   Trends in space

Are there geographic patterns in macroevolutionary processes? The evidence suggests so. Jablonski (1993) showed that the first occurrence of marine invertebrate fossil groups was more often in the tropics than would be expected by chance. This points to a higher origination rates of higher taxa there, perhaps due to fostering of important morphological innovations that could give rise to radiations. These innovations also occur in shallow water environments more often than deep water environments. Stehli *et al.* (1972) also showed that extant species of **Foraminifera** do not vary in age according to their present day latitudes. This suggests that extinction rates do not vary with latitude, and consequently invoking higher speciation rates in the tropics to explain their higher diversity there. In contrast, the ages of higher taxa of corals, bivalves,

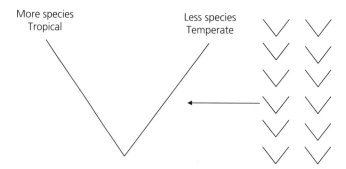

More species
Tropical

Less species
Temperate

**Fig. 14.5** Correlating diversification rates with latitude. Sister clades, sharing a common ancestor and being the same age, are compared. One of the clades is largely tropical, the other temperate. The difference in their species richness is also compared. If over a number of such pairs of sister clades, the more tropical clade has the most species, then latitude affects the net rate of diversification.

and barnacles decline towards the tropics. This suggests that speciation rates are higher there (Sepkoski 1999). Studies of extant taxa also suggest similar trends. Comparisons of the species richness of tropical and temperate sister taxa in swallowtail butterflies and passerine birds suggests that the net rate of diversification has been higher in the tropics (Cardillo 1999) (Figure 14.5). The average age of avian tribes per latitudinal band increases towards the tropics (Gaston and Blackburn 1996), suggesting a loss of older tribes towards higher latitudes. In contrast, northern hemisphere freshwater fish species show a steep decline in diversity north of 50 degrees, in areas recently glaciated. Interestingly, however, the mitochondrial DNA distance between sister taxa is lower at these high latitudes, suggesting a recent rise in speciation rate with latitude following colonization of these lakes, a trend consistent with the notion of ecological opportunity (Bernatchez and Wilson 1998).

Finally, reconstructing the history of geographic ranges in groups from phylogenies of living species has suggested that some radiations have originated in the tropics (Bleiweiss 1998, Böhm and Mayhew 2005 ). What has happened to morphology in these communities? A few studies have compared the morphospace occupied by temperate and tropical communities. Ricklefs and O'Rourke (1975) found that disparity increased in tropical moth communities, with new morphologies not found in temperate communities. Similar trends have been found in tropical marine communities (Roy and Foote 1997).

## 14.4    Trends across taxa

What of differences between taxa? There are consistent messages from both fossil studies and studies of extant taxa. Across taxa, speciation rates and extinction rates tend to be positively correlated (Stanley 1979; Sepkoski

**Fig. 14.6**    A weevil, family Curculionidae, walking on a pencil. Weevils are the most species-rich family, in the most species-rich order, class, phylum, and kingdom. Photo © Peter Mayhew.

1999). One possible reason is that extinction and speciation are both promoted by the same characteristics. Data from Jablonski (1986) on North American Late Cretaceous gastropods illustrate a potential mechanism: those taxa which have planktonic larvae, and hence are good dispersers, have wide geographic ranges, and experience low extinction rates but also low speciation rates. In contrast, taxa with direct development (no planktonic stage) have small geographic ranges and tend to have high extinction rates, but also high speciation rates. In many cases, however, we can identify taxa as differing in rates of some cladogenetic process without knowing the reason because the taxa are unreplicated; any feature that differs between the two taxa could be responsible. For example, of the four major primate groups (lemurs and their kin, new world monkeys, old world monkeys, and **homi-noids**), only the old world monkeys differs from the others in net rate of diversification, and furthermore, this is due to a higher rate of speciation rather than lower rate of extinction (Purvis *et al.* 1995). This can be inferred from the relative shapes of the phylogenies of the groups (Figure 14.2), which for the old world monkeys shows a recent burst of branching indicating a high rate of speciation over extinction. Similarly, based on their ages and extant species richness, the beetles (Figure 14.6) show a higher net rate of diversification than their sister group (Mayhew 2002), though there are many possible reasons. Other such 'significant radiations' include the canids (Bininda-Emonds *et al.* 1999), and the passerine and wading birds (Figure 14.7) (Harvey *et al.* 1991).

**Fig. 14.7**    A snipe, *Gallinago gallinago*, a member of the order Charadriiformes; one of two bird orders that are significant radiations. Photo courtesy of Stephane Moniotte.

One way of pinning down the reasons for differences in rates of cladogenesis is if replicate radiations have occurred in different groups with similar characteristics (Figure 14.5). One can then compare the species richness of these groups with those of their sister taxa which lack the characteristic or 'key innovation' in question. A large number of such studies have now been done. A number of these are consistent with theory of speciation and adaptive radition. For example, promiscuous insect groups have diversified more rapidly than their sister groups with other mating systems (Arnqvist *et al.* 2000). Promiscuous insects are those in which a single female is mated by many males, and the genetic interests of male and female are predicted to be divergent (see Chapter 7), leading to rapid evolution of sexual traits in an intraspecific arms race. This rapid evolution could lead to rapid evolution of reproductive isolation. Another trait that is linked with elevated species richness is sexual selection. In passerine birds, a higher incidence of sexually dimorphic coloration, indicative of sexual selection, is associated with elevated species richness (Barraclough *et al.* 1995). Nectar spurs in plants may have a similar effect by association with specific pollinators which promote reproductive isolation (Hodges and Arnold 1995). Other traits seem more likely to be linked with ecological opportunity. The link between species richness and phytophagy, for example, seems likely to be related to this (Mitter *et al.* 1988; Farrell 1998), as does latex and resin canals in plants, which provide defence against herbivores (Farrell *et al.* 1991) allowing 'escape and radiation' (Chapter 11).

To get at extinction correlates we can also draw on the sobering recent dataset on anthropogenic extinction. How do species that have recently gone extinct differ from their close relatives that have not? In a number of groups, correlates of extinction risk also match the theory of extinction (Fisher and Owens 2004) (Chapter 13). In primates, carnivores and birds, small geographic range is correlated with extinction risk, as is dietary or habitat specialization in hoverflies, reptiles, birds, and primates, and low population density in reptiles, carnivores, and primates. Large body size gives a higher risk of extinction in birds, primates, and hoverflies.

There are few surprises here. What is particularly interesting, however, is the interaction between the source of extinction threat and traits in birds. Birds that are small are likely to be at risk from exploitation or introductions, but not from habitat destruction. However, birds with specialized habitats are more at risk of extinction from habitat destruction, but not from exploitation or introductions (Owens and Bennett 2000). These threat interactions may account for some of the contradictory results obtained from some studies of extinction. For example, in mammals species in taxonomically isolated groups are most at risk of extinction whereas in angiosperms it is species in species-rich groups. Body size has been reported to have positive effects on extinction risk in primates and birds, but negative effects on all mammals, and no effect in a large number of other studies. Generation time has positive effects in carnivores, negative effects on birds, and no effects on primates, reptiles, and hoverflies (Purvis *et al.* 2000).

The combination of fossil evidence and data from extant taxa then is starting to make headway into the dynamics of clade characteristics and differences between them. Taxonomic diversity and morphological diversity are both promoted by ecological opportunity but they may not follow directly parallel paths. In addition a number of characteristics affect speciation and extinction rates that match the predictions of speciation and extinction theory. In the next chapter we will examine large-scale patterns from a more ecological perspective.

# 14.5   Further reading

Signor (1990) reviews the fossil evidence for trends in time, and Sepkoski (1999) does the same across taxa. Purvis (1996), Barraclough *et al.* (1999), Harvey *et al.* (1991), and Nee *et al.* (1996) review the use of phylogenies in macroevolution. Purvis *et al.* (2000) and Fisher and Owens (2004) review trends in extinction across taxa. Foote (1997) and Roy and Foote (1997) review disparity.

# 15 Big ecology

Nature uses only the longest threads to weave her patterns, so that each small piece of her fabric reveals the organization of the entire tapestry.

Richard P. Feynman

Macroecology is the field that describes and attempts to explain statistical patterns of abundance, distribution, and diversity (Brown 1995; Gaston and Blackburn 2000). Macroecologists have identified a number of distinctive patterns in these variables that seem to be telling us something important about the rules governing ecology and evolution. In the development of the field, identifying patterns constitutes what Gaston and Blackburn (1999) describe as the 'what' stage of development, and is reasonably well advanced. Ultimately, we would like to know 'how' and 'why' the patterns are generated. This is not only because the patterns themselves seem to be important features of our biotic universe, but also because underlying them are fundamental principles of the workings of ecology and evolution. We are groping for the rules behind a great ecological and evolutionary game, perhaps the greatest of them all. Since by far the most effort so far has gone into documenting the patterns, the field is very much dominated by empirical data and not by theory. Nonetheless, many hypotheses have now been formulated to explain many of the patterns, and some data allow comment on the merits of these hypotheses. One particular obstruction is that the nature of the patterns restricts tests of the hypotheses. One particular empirical option, experiment, is largely ruled out by the practical concerns of scale. As we shall see, however, this does not exclude useful tests that can distinguish between alternatives.

Several macroecological patterns rightly find prominence in ecology texts but rather fewer find their way into evolutionary texts. This is probably a mistake because, as I hope to show below, many of the patterns require us to consider evolutionary processes and mechanisms as well as ecological ones. Testing hypotheses about these mechanisms will therefore be aided by evolutionary theory and data. Before discussing some of the theory and data, however, let us outline some of the patterns we are trying to explain. The patterns fall into two categories: associations between variables and frequency

**Table 15.1**  Some of the best known macroecological patterns

| Type of pattern | Variables concerned | Nature of pattern | Ubiquity of pattern |
|---|---|---|---|
| Frequency distribution | Latitude of occurrence | More species found at low latitudes than high latitudes | Virtually ubiquitous except at very small scales and taxa |
| | Geographic range size | Bimodal or right skewed and unimodal | Bimodal at small scales, unimodal and right skewed at large scales |
| | Body size | A right skewed distribution with many more small than large species | Fairly ubiquitous especially at large spatial and taxonomic scales |
| | Abundance | Most species rare | Virtually ubiquitous across scales and taxa |
| Associations between variables | Body size and latitude | Larger bodied species at higher latitudes (Bergmann's rule) | Not ubiquitous but known from several bird assemblages |
| | Body size and abundance | Negative relationship | Very common, but found less frequently at smaller scales |
| | Geographic range and latitude | Larger geographic ranges at high latitudes (Rapoport's rule) | Not ubiquitous but known from several bird assemblages in the Holarctic |
| | Abundance and geographic range | More widespread species achieve higher local abundance | Virtually ubiquitous across scales and taxa |
| | Species richness and geographic area | Positive relationship | Virtually ubiquitous across scales and taxa |

distributions of single variables (Table 15.1). Each relationship can be examined across a range of taxa, of different rank, and at a variety of spatial scales. Spatial scale can mean the total area over which the relationship is examined, and/or the area that comprises an individual data point (quadrat size). When relationships are examined over large geographic areas, quadrats naturally tend also to be large. For some relationships, data are relatively taxonomically restricted; for example, geographic range is only well documented in a large number of species for a few higher taxa, mostly birds and mammals. In contrast, data on local abundance or species richness come from numerous studies on many taxa. Patterns are mostly examined over quite large taxonomic ranks, such as classes. Changing the taxonomic rank under consideration, may, as we shall see below, affect the pattern. Putting together frequency distributions is relatively straightforward and requires few special considerations. However, relationships between two variables often means comparative analyses of species characteristics, and then phylogenetically based comparative techniques should be used where possible (Chapter 4). In the remainder of the chapter, we will consider two of the most robust patterns and how evolutionary processes may contribute to them.

## 15.1 The evolution of frequency distributions in species body sizes

Most species contributing to a large taxonomic rank are small. This appears to be the overwhelming message from an examination of species body size distributions (Figure 15.1). The distributions tend to be right skewed, even on a logarithmic scale, and yet, the mode is seldom in the smallest size class but slightly larger. These patterns probably hold regardless of potential biases in the data (Blackburn and Gaston 1994). The skewness tends to be less common at smaller spatial scales and when small taxonomic ranks (i.e. families instead of classes) are examined (Maurer 1998a).

What processes might account for this pattern? Dial and Marzluff (1988) and later McKinnley (1990) both developed verbal/graphical evolutionary models based on evolutionary mechanisms. Dial and Marzluff assumed that the difference between speciation and extinction (net diversification) is greater at small than large body sizes (Figure 15.2). Then one can imagine smaller-bodied lineages on the left-hand size of the frequency distribution out-proliferating those on the right. McKinnley developed a passive

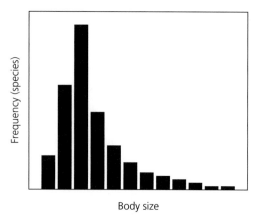

Frequency (species)

Body size

**Fig. 15.1** A typical frequency distribution of species body sizes. The distribution is right skewed, with the mode well to the left, but not at the smallest size class.

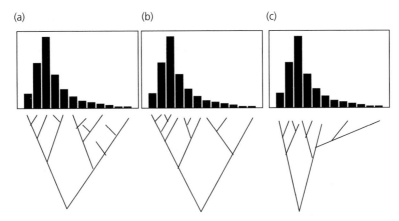

**Fig. 15.2**   Possible evolutionary pathways to the right-skewed frequency distribution of species body sizes. (a) The ancestor was medium sized, and speciation occurs with equal frequency at all body sizes, but larger species go extinct more rapidly. (b), Speciation occurs more frequently at small body sizes. (c) The ancestral body size was small, and only rarely do some species become large, hence anagenesis is biased.

diffusion scenario of body size evolution. He assumed first that clades originate at small body sizes, and that there is a lower limit to body size. Anagenetic processes must also play a role if a clade is to produce variable body sizes; this might occur directionally (towards larger body sizes) or in a random direction in a process akin to drift. As a result, the clade multiplies fastest at smaller body sizes and develops a right-hand tail (Figure 15.2).

These ideas were first formalized in simulation studies of clade growth and evolution by Maurer *et al.* (1992), whose approach was developed still further by Kozlowski and Gawelczyk (2002). The simulation models contain the following variables, which were examined to observe if they can produce the shape and range of observed patterns. First, the body size of the clade founder; second, the size-dependence of speciation rate; third, the size-dependence of extinction rate, and fourth, the presence of a 'reflecting barrier' at small body sizes. The latter represents a size limit below which species cannot, or find it difficult to persist due to physiological constraints. Kozlowski and Gawelczyk (2002) made this a gradual constraint such that anagenesis to smaller body sizes becomes increasingly less likely below a certain size. Fifth, body size change can occur either at speciation events, or between them (anagenesis), although the direction of change was assumed to conform to a random drift process (i.e. no trend).

The results of the models are mostly intuitive. The presence of the graded reflecting barrier produces the short left-hand tail rather than a truncated distribution. The simulations produce a range of skewness as seen in nature due to the stochastic nature of all the processes assumed. When the net diversification rate favours small species, more right skewed distributions are produced and they tend to become more heavily right skewed. When the reflecting barrier is present, small founder body size tends to increase skewness and proportion of skew, although without it the opposite is the case. The timing of body size change affects the relative importance of size-biased extinction and speciation.

In general then, these evolutionary processes are capable of producing the distributions seen in nature and certain combinations are more likely to do so than others. For expiricists the challenge is now to quantify them in nature and compare to theoretical predictions. Most work on this has been directed at the macroevolutionary processes of speciation, extinction, and net cladogenesis. Orme *et al.* (2002) have studied size-related trends in net cladogenesis by observing the correlation between body size and species richness across 38 species-level phylogenies across range of vertebrates and invertebrates. Unfortunately, the results pose more questions than they answer. There is no overall trend towards an association between body size and net rate of cladogenesis, and in fact only one study does show this—a genus of flies (*Bitheca*), where small body size increases the net rate of cladogenesis. Furthermore, there was no relationship between body size and distance from the root of the tree, so it appears that there are no strong and consistent anagenetic trends. These do sometimes appear, however, in particular clades; significant relationships are found in eight clades; in primates, body size increases with distance from the root, while in carnivores it decreases; overall, three of these cases show a positive relationship and five are negative.

These results do seem at odds with the evolutionary models, and there are several possible explanations; one is that there is actually no consistent skew in the body size distributions of those taxa examined. If this is the case, the studies can hardly be considered as explanations for positive skew. However, the studies are also all much smaller than those generally used to describe the patterns of interest, and, as described earlier, taxa of small taxonomic rank tend to display less skewed body size distributions. It may be therefore that we need to wait for larger phylogenies to become available. An alternative possibility, not considered in the modelling studies, however, is that small size is not a consistent correlate of species diversity but that large radiations might tend to be small bodied. In fact this pattern has been documented in mammalian clades by Gardezi and da Silva (1999). Such radiations would have large effects on species richness but represent only single data points so are unlikely to affect a general bias in cladogenesis. Another suite of studies have examined the relationship between present day extinction risk and body size in a number of groups. They generally conform to the studies of cladogenesis in that some positive relationships are found, some negative

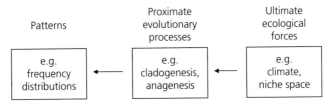

**Fig. 15.3**   Processes contributing towards macroecological patterns at very large (e.g. global) scales. The patterns are the direct result of evolutionary processes, which are thus proximate forces. These forces occur in the way they do because of ecological forces, which are thus ultimate.

ones, and in some there is no relationship (Purvis *et al.* 2000). In general then, there is presently scant evidence for any of the postulated evolutionary processes that might have given rise to body size distribution patterns; the reason for the mismatch must be found.

Even if we had a good knowledge of the evolutionary processes mentioned above, that would remain an incomplete explanation, for underlying these must themselves be causal processes (Figure 15.3). Three major ultimate reasons have been postulated. First, there might be more resources or niches available for smaller bodied species that might increase rates of cladogenesis or affect anagenetic trends. In particular one can describe the relationship between body size and number of individuals from measurements of the habitat at different scales. Some studies of arthropod communities (Morse *et al.* 1985; Shorrocks *et al.* 1991) show reasonably good fit between the observed and predicted number of individuals. Of course the link between amount of resource/niche space and evolutionary process remains rather implicit. A second postulated ultimate process is that organisms might be evolving towards some theoretical optimum size for the clade based on reproductive power (Brown *et al.* 1993). The idea here is that within a clade there will be some characteristic body size at which most energy becomes available for reproduction, due to the differential between the mass specific gains and losses of energy. Brown *et al.* (1993) predict that the relationship between production rate (giving reproductive power) and body size is shaped rather like species body size distributions. Other factors not in the model can combine to drive a species away from the optimum and this is more likely when the reproductive power is close to the maximum achievable. Thus, the species distribution comes to resemble the body size–production relationship.

The Brown *et al.* (1993) model apparently predicts the modes of some body size distributions quite well (Maurer 1998; Roy *et al.* 2000). However, these tests have themselves been criticized on the grounds of loose application of the necessary parameters (Kozlowski 2002). In addition the model itself has been criticized on a number of mathematical issues, points of internal consistency, and on biological and evolutionary grounds (for a review of these points see Kozlowski 2002). In addition two further tests of the model, both of which look for the proposed switch point in reproductive power at the modal body mass, fail to confirm the model prediction (Jones and Purvis 1997; Symonds 1999). Once again, it is rather implicit which evolutionary processes actually lead to the species body size distribution. Presumably anagenetic change is involved, as the model is really about what an individual species' body size should be. Presumably though, how many species get close to that could be affected by speciation or extinction rates, which might themselves be affected by the relatively reduced reproductive power away from the optimum.

An alternative kind of optimality approach has been developed by Kozlowski and Weiner (1997), in a model that developed Charnov's (1991) model of mammalian life history evolution. In Chapter 4, we saw how the trade-off between production and mortality could combine to influence when a species should mature, hence, its optimum adult body size. Kozlowski and Weiner (1997) investigated a slightly more complex model than Charnov's in that they derive production from separate allometric relationships of assimilation and respiration. Varying the parameters of this model by drawing them at random from normal distributions happens to produce distributions of optimal species body sizes that are right skewed and shaped very like those in nature (see also Kindlmann *et al.* 1999; Kozlowski and Gawelczyk 2002). Exactly how this optimization process affects the required evolutionary processes is again implicit; presumably ecological conditions make some parameters more common than others, our third ultimate process, and this affects the frequency of optima and hence of anagenetic change towards those optima. Not all organisms optimize body size in this way (parasitoids and birds are obvious exceptions—see Chapter 4), but the model illustrates the general principle of how evolution on a *per species* basis towards individually determined optima can lead to species body size distributions.

We will now turn to another well-known pattern and see how the same sets of evolutionary principles can help explain that.

## 15.2   Evolution of the latitudinal gradient in species richness

Following on from the last chapter, we can refer to the same suite of evolutionary processes to explain why, at larger taxonomic and spatial scales, species richness increases as latitude declines (Figure 15.4). Remarkably, the latitudinal gradients are seldom considered in these terms, yet logically, they are likely to be formed by the same suite of processes. Clades originate somewhere and, all other processes being equal, their centre of species richness will remain

(a)                                                (b)

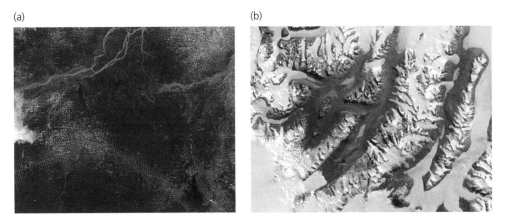

**Fig. 15.4**   The latitudinal gradient in species richness seen from space. (a) Amazonia, with many species. Deforestation along a road is visible in the bottom half of the picture. (b) Antarctica (the McMurdo dry valleys) with very few species. Photos from the NASA Visible Earth image archive.

there. We must also consider how speciation and extinction vary with latitude and the nature and frequency of latitudinal shifts in geographic range (anagenesis). Just as when body size distributions could be considered to be bounded by a lower reflecting barrier, so geographical limits are placed on latitudinal range shifts, provided by the positions of land masses and oceans, for example, and on the fact that the latitude cannot vary below 0° or above 90° from the equator. Some of these geographical limits have been assessed in a number of non-evolutionary models, and have been criticized exactly because of their non-evolutionary nature. In fact the suite of evolutionary processes that I have just described are the well-known processes of evolutionary biogeography, and in a proximate sense explaining the latitudinal gradient is an exercise in evolutionary biogeography. We would expect that biased originations close to the equator as well as increased rates of cladogenesis, reduced rates of extinction, and biased shifts in range towards the equator would all contribute towards the observed latitudinal gradient (Figure 15.5).

Once again some progress has been made in documenting these processes. In Chapter 14, we saw that the fossil record suggests higher origination rates in the tropics, and sister-taxon comparisons of species richness suggest net rates of cladogenesis that are sometimes higher in the tropics. Taxon age also seems to vary with latitude in many instances, suggesting variation in underlying processes. In baboons, macaques, and their relatives (the Papionini), the history of range changes has been reconstructed by mapping extant distributions onto a phylogeny of species. These reconstructions show, robustly, that ancestral distributions were equatorial, that tropical regions have experienced more net cladogenesis, and that tropical regions have given

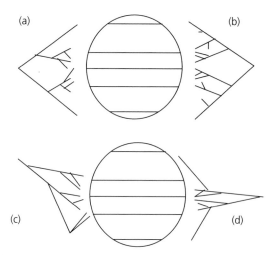

**Fig. 15.5** Four ways in which evolutionary processes may have given rise to the latitudinal gradient in species richness (the circle with horizontal lines indicates the Earth with lines of latitude). (a) Speciation is more frequent in the tropics. (b) Extinction is more frequent near the poles. (c) Species move towards the equator over time. (d) The ancestor was tropical and changes in range towards higher latitudes are rare.

rise to more dispersal, suggesting that they actually raise the species richness of other regions (Böhm and Mayhew 2005).

In addition to documenting these processes, we will also need to find the fundamental reasons behind them; ultimate processes, and once again we will therein need to invoke ecological characteristics that have influenced them. Here there have been so many different hypotheses that I cannot do justice to them all, so I will mention one recent theory that makes quantitative predictions, and has some empirical support (Allen *et al.* 2002). First, assume that the total energy flux per species per unit area is relatively invariant (the energetic equivalence rule). This is an empirically derived rule that appears to hold and is derivable for some plants based on scaling relationships (Chapter 4).

Now assume that in a community of individuals made up of different species of similar type and ecology (e.g. forest trees, reptiles, beetles), the total number of individuals is determined by the number of species and the number of individuals per species. If the energetic equivalence rule is to hold then, at higher ambient temperatures, to maintain the same temperature flux per species there must be fewer individuals per species. If there are fewer individuals per species and the total number of individuals in the community is approximately invariant with respect to temperature, then the number of species must increase. This logic actually allows quantitative prediction of the relationship between species number and temperature for ectothermic organisms once the exact metabolic relationships are taken into account. The exact prediction is that a plot of ln species richness against

1000/K should give a slope of −9. A number of tree, amphibian, fish, gastropod, and parasite communities conform to this relationship across both latitudinal and elevational gradients.

The model of Allen *et al.* (2002) is remarkable in several ways; first, it makes quantitative predictions which can be more easily falsified than many other models. Second, it derives species richness directly from a primary environmental variable that is directly related to latitude, but actually varies in a much broader sense. It is not specific about the evolutionary processes but presumably the energetic equivalence rules could affect the likelihood of range shifts, speciation, and extinction rates primarily by allowing species to fit into a community in energetic terms or not. In many ways, the evolutionary processes can be considered to be irrelevant because as long as they produce enough species to fill up the energetic opportunities, the model will hold. It is therefore a theory based largely on ecological equilibrium, with some implicit evolutionary assumptions. Because of its novelty it has yet to be fully assessed by the scientific community, so I will tempt fate by pointing out two likely areas of criticism. The first is that it cannot predict the species richness of endotherms, because, using the same logic, their metabolic rates are invariant with ambient temperature, the number of individuals per species does not change with temperature, and the number of species should therefore remain the same. The second likely area of criticism is whether most ecological communities are at energetic equilibrium. In general, this goes counter to most ecologists' notions of community assembly.

Thus, many macroecological patterns can be derived from consideration of proximate evolutionary processes many of which have begun to be quantified. Ultimately, these processes rely on the interplay between primary ecological phenomena and the various proximate evolutionary processes. Some possible primary phenomena have been identified and some data supports their action, but the link between the ecological and evolutionary phenomena remains poorly described theoretically and totally undescribed empirically. There is much to be done to develop our understanding of how and why these ecological patterns emerge and it is essential that evolutionary biologists play a leading role.

## 15.3   Further reading

Brown (1995) and Gaston and Blackburn (2000) review macroecology, and some recent reviews are in Blackburn and Gaston (2003). Kozlowski and Gawelczyk (2002) review body size evolution, and Orme *et al.* (2002) cover the phylogenetic evidence. Willig *et al.* (2003) provide an overview of the latitudinal gradient in species richness from a statistical perspective.

# 16 Combining in diversity

Coming together is a beginning; keeping together is progress; working together is success.

Henry Ford

Over the previous 14 chapters, I have recounted the major questions that evolutionary ecologists have addressed and some of the discoveries that they have brought to light. In this chapter, I will draw on this experience to explain what evolutionary ecology as a whole is trying to do. First, evolutionary ecology is primarily about understanding biological diversity. Both ecology and evolution are about understanding variation in a common set of variables, so it is natural that they should sometimes need to work together to do that. Second, evolution and ecology affect each other. Evolutionary ecology explores the different ways in which this happens. Third, the different topics in evolutionary ecology help explain each other. This occurs because many evolutionary or ecological traits are dependent on each other, and because many of the research tools are broadly applicable. Below I expand briefly on these themes in the light of the preceding chapters.

## 16.1 Understanding biological diversity

I have stated that evolutionary ecology is collectively about understanding diversity. Biological diversity is expressed through variation in the characteristics of individuals, populations, communities, and clades. The characteristics that differ include phenotypic traits, population size, and species richness (Table 16.1). How has evolutionary ecology addressed these characteristics?

**Table 16.1** The entities and traits that ecologists and evolutionary biologists study

| Entity | Individual | Population | Community | Clade |
|--------|-----------|------------|-----------|-------|
| Trait | Phenotypes | Number of individuals | Number of species | Number of phenotypes or species |

Chapters 2 and 3 assessed how major innovations in phenotypic characteristics occurred. These changes increased the total diversity of traits on the planet, and many were successful in a macroevolutionary sense, increasing species richness or other community properties. They explain why the world is a rich and complex place. Chapters 4 to 7 considered how variation in phenotype arises even without major innovations discussed in earlier chapters. They explain the little traits that make even closely related species or individuals differ in form and function. In Chapter 8, we considered how population size can be controlled through evolutionary processes. Rather interestingly, we can not only explain variation in population size and dynamics through trait evolution, but also trait evolution through population dynamics. The result is not just variation in numbers but also variation in form and function.

In Chapters 9 to 11, we considered properties of species that are also properties of communities: their interactions with other species. Ecological interactions evolve between generalism and specialism, between mutualism and antagonism, or can follow a large number of co-evolutionary pathways, and because of that the world is full of complex ecological communities and of species that differ. In Chapters 12 to 15, we considered properties of communities and clades, such as their species richness and constituent body sizes. The rates of change in cladogenesis and morphological variation differ over time and space and among lineages, and in doing so create the patterns in traits and species richness that dominate our world. In these ways evolutionary ecology addresses some of the major questions posed by ecology and evolution, and also some of the major questions of any kind about the biosphere and its constituents.

## 16.2   How ecology and evolution interact

The previous chapters also illustrated how ecology and evolution affect each other reciprocally and in diverse ways (Table 16.2). Throughout the book a prominent issue has been how ecology shapes anagenetic evolution through natural selection. In Chapters 4 to 7 we considered how the day-to-day phenotypic variation among species evolves through selection pressures exerted by their environment. These include changes in life history variables, such as reproductive lifespan and age at maturity, changes in allocation of reproductive effort to male verses female function and in the way sex is determined, changes in dispersal ability and dormancy, and plastic changes in phenotype exerted during the lifetime of individuals in response to changes in their environment. In each of these cases, a consideration of the fitness consequences of alternative phenotypes has led us to understand the circumstances that would favour the genotypes that code for them.

**Table 16.2** How ecology and evolution affect each other

| Influencing factor | Influenced factor | Reason |
|---|---|---|
| Ecology | Phenotypic change | Natural selection |
| Ecology | Speciation | Reproductive isolation, ecological divergence |
| Ecology | Extinction | Reduction in range or population size |
| Evolution | Ecology of individuals | Natural selection |
| Evolution | Population ecology | Rapid evolution or adaptive behaviour |
| Evolution | Community ecology | Selection on species interactions, changes in species richness |
| Evolution | Ecosystem ecology | Evolutionary novelties affect geochemical cycling, ecospace occupation |

Ecology also affects cladogenesis through its impact on speciation. Speciation involves reproductive isolation and ecological divergence, and ecology is involved in both. Reproductive isolation results from the action of selection or other ecological forces, such as drift, that require specific ecological conditions, and are often a consequence of geographic isolation. Speciation can also occur through hybridization, requiring proximity of two close relatives. Ecological divergence occurs as a result of selection, or neutral processes that require specific ecological conditions.

Ecology also affects evolution through its impact on extinction. Extinction occurs when ecological forces combine to make the range or abundance of a species small, after which a range of stochastic and deterministic events can lead to irretrievable loss of fitness. Ecology also drives the diversity of form and species richness among clades through competition and ecological release.

Evolution also drives ecology. The ecology of individuals is affected by plastic responses to changes in their environment. Evolutionary forces have shaped these such that they are adaptive. Evolutionary forces, acting through these plastic changes, can also affect how the size of a population will change in response to changes in the environment. Rapid evolution can also affect such changes. In addition, evolutionary forces can make a species rare or can make it go extinct once rare. Evolution can also affect the properties of communities: it can affect the number and type of species interactions, making some species generalist and others specialist, some relationships antagonistic and others mutualistic. In general, there has been an increase in the complexity of earth's communities and ecosystems over time, affected by what might be termed major transitions in ecology. Evolutionary forces have shaped those changes through stepwise addition of small evolutionary changes, each of which was advantageous in the ecological circumstances of the organisms concerned. Some of those changes involved an increase in species richness, another property of communites. In fact anagenesis and cladogenesis have affected many of the key macroecological patterns that today dominate the living world.

Thus, evolutionary ecology has so far uncovered a rich array of interactions between ecology and evolution, and knowledge of those interactions has been the key to answering some important questions about our planet.

## 16.3   The interaction of questions in evolutionary ecology

Throughout the book several issues have been raised in more than one chapter, implying that understanding one topic in evolutionary ecology can enhance understanding of another. This occurs for two reasons. First, the topics are biologically interdependent. Second, the techniques we can use to aid understanding are often general. These connections have also led many notable evolutionary ecologists to move from one subject to the next. Let us now recap the connections illustrated by the book (Figure 16.1).

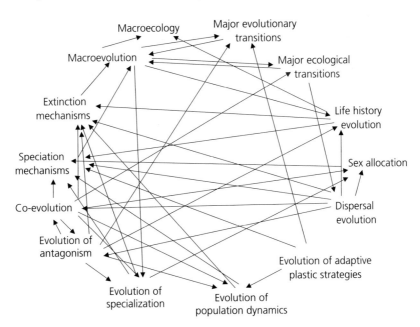

**Fig. 16.1**   The interactions between different subject areas in evolutionary ecology mentioned in this book.

In the last chapter, we saw that understanding macroecological patterns is sometimes helped by an understanding of life history evolution. For example, one macroecological pattern, the body size frequency distribution, involves variation in a life history trait. Macroevolutionary forces, such as rates of speciation and extinction, also affect many of these patterns, such as why there are more species in the tropics. Macroevolutionary patterns are also sometimes the result of major ecological or evolutionary transitions, such as the evolution of sex or flight. They are often explicable by reference to speciation and extinction theory, such as in the haplochromine cichlids of Lake Victoria. Extinction is often the result of variation in life histories, such as fecundity, ecological specialization, with specialists being more extinction prone, or co-evolutionary forces, such that invasive species can cause extinction of species that they have not co-evolved with. The theory of extinction suggests that species that have small ranges are at risk of extinction, and this explains some macroevolutionary observations on extinction rates across taxa.

Speciation can be enhanced by rapid evolutionary forces, such as sexual selection, as in haplochromine cichlids, or conflict between evolutionary entities, as in cytoplasmic male sterility, or changes in ecological specialization, as in apple maggot fly, or co-evolutionary forces, as in diversifying co-evolution. Co-evolution characteristically results in changes in antagonism, changes in specialization, changes in population dynamics. Changes in antagonism have resulted in many of the major evolutionary transitions, can affect specialization through expansion or contraction of the ecological niche, can affect life history evolution through trade-offs with virulence, population dynamics through its influence on vital rates, and can result in extinction as a co-evolutionary process. Evolutionary changes in population dynamics can be caused by adaptive changes in behaviour, as in many bird and mammal species, can cause speciation through evolutionary branching and extinction through adaptive suicide.

Plastic phenotypic responses can cause speciation, such as through mate choice or conflict between entities. Dispersal and dormancy affect the degree of antagonism between species and among species, and sex allocation through determining population structure, and life history evolution through trade-offs, and speciation, and extinction rates. Sex allocation is affected by co-evolution and conflict of interest among entities and can cause speciation. The major transitions in ecology are recognized by their macroevolutionary effects, and some, such as the evolution of flowers and of animal phyla, may have been the result of co-evolutionary forces. Others, such as flight, have affected dispersal ability in many taxa.

Are any of these topics particularly pervasive? Cooperation and conflict, dispersal, and life history evolution permeate many areas of evolutionary ecology and might be considered central. Macroevolution is a force contributing to phenotypic distribution of many of the traits of interest.

In addition to this web of interactions between the topics is a web of mutually useful methods and tools. Optimization theory, evolutionary stable strategies, adaptive dynamics, and population genetics are theoretical tools that are broadly applicable in evolutionary ecology. Similarly, empirical tools such as phylogenies and the comparative method, unmanipulated observations, and controlled experiments are useful in nearly all areas. Thus, the tools of evolutionary ecology as well as the concepts are mutually supporting.

The picture provided by Figure 16.1 is a picture of evolutionary ecology, it also describes an evolving biosphere and the forces that shape it. It is the message of this book.

# 16.4   Prospects

While evolutionary ecologists have come a long way, there is obviously much still to do. We still have a very sketchy empirical knowledge of many topics, such as co-evolution, the evolution of dispersal, most of the early major transitions in ecology and evolution, macroevolution and even macroecological trends for most taxa. Some areas are also theoretically poor, in some cases due to a lack of data; macroecology and macroevolution notably so. Some topics are recent and will inevitably see big advances in knowledge and direction; the evolution of population dynamics is one such area. Other areas are mature in data and theory, and yet are still experiencing interesting new changes, such as in sex allocation and life histories. Substantial progress in understanding is therefore expected in all areas. Because of the interconnections between the different topics, we would expect progress in one area to continue to advance others. Much of the progress will come through innovative new techniques, both theoretical, like adaptive dynamics, and empirical, like molecular phylogenetics. However, evolutionary ecology has always gained much, and will continue to gain much, by simple work that asks the right questions.

Evolutionary ecology is wonderful for what it gives us and what it demands of us. It gives us understanding of the natural world that we have always striven for. It demands of us our best characteristics: ability, imagination, energy, and love. As contributors towards society and its progression, we all deserve to appreciate and celebrate it.

# References

Abrams, PA (2001). Predator–prey interactions. In CW Fox, DA Roff, and DJ Fairburn, eds. *Evolutionary ecology: concepts and case studies*, pp. 277–289. Oxford University Press, New York.

Allen, AP, Brown, JH, and Gillooly, JF (2002). Global biodiversity, biochemical kinetics, and the energy equivalence rule. *Science*, **297**, 1545–1548.

Allender, CJ, Seehausen, O, Knight, ME, Turner, GF, and Maclean, N (2003). Divergent selection during speciation of Lake Malawi cichlid fish inferred from parallel radiations in nuptial coloration. *Proceedings of the National Academy of Sciences, USA*, **100**, 14074–14079.

Anderson, RM and May, RM (1982). Coevolution of hosts and parasites. *Parasitology*, **85**, 411–426.

Arnold, ML (1997). *Natural hybridization and evolution*. Oxford University Press, Oxford.

—— and Emms, SK (1998). Paradigm lost: natural hybridization and evolutionary innovations. In DJ Howard and SH Berlocher, eds. *Endless forms: species and speciation*, pp. 379–389. Oxford University Press, New York.

Arnqvist, G, Edvardsson, M, Friberg, U, and Nilsson, T (2000). Sexual conflict promotes speciation in insects. *Proceedings of the National Academy of Sciences, USA*, **97**, 10460–10464.

Atlan, A (1992). Sex allocation in an hermaphroditic plant: the case of gynodioecy in *Thymus vulgaris*. *Journal of Evolutionary Biology*, **5**, 189–203.

Barton, NH (1998). Natural selection and random genetic drift as causes of evolution on islands. In PR Grant, ed. *Evolution on islands*, pp. 102–123. Oxford University Press, Oxford.

Barraclough, TG and Vogler, AP (2000). Detecting the geographical pattern of speciation from species-level phylogenies. *American Naturalist*, **155**, 419–434.

—— Harvey, PH, and Nee, S (1995). Sexual selection and taxonomic diversity in passerine birds. *Proceedings of the Royal Society of London, Series B*, **259**, 211–215.

—— Vogler, AP, and Harvey, PH (1999). Revealing the factors that promote speciation. In AE Magurran and RM May, eds. *Evolution of biological diversity*, pp. 202–219. Oxford University Press, Oxford.

Bawa, KW (1980). Evolution of dioecy in flowering plants. *Annual Review of Ecology and Systematics*, **11**, 15–39.

Bell, G (1982). *The masterpiece of nature: the evolution and genetics of sexuality*. Croom Helm, London.

Benkman, CW (1999). The selection mosaic and diversifying coevolution between crossbills and lodgepole pine. *American Naturalist*, **153** (Suppl.), S75–S91.

—— (2003). Divergent selection drives the adaptive radiation of crossbills. *Evolution*, **57**, 1176–1181.

Benner, SA, Ellington, AD and Tauer, A (1989). Modern metabolism as a palimpsest of the RNA world. *Proceedings of the National Academy of Sciences, USA*, **86**, 7054–7058.

Bennett, PM and Owens, IPF (2002). *Evolutionary ecology of birds: life histories, mating systems and extinction*. Oxford University Press, Oxford.

Benton, MJ (1995). Diversification and extinction in the history of life. *Science*, **268**, 52–58.

Berenbaum, MR (1983). Coumarins and caterpillars: a case for coevolution. *Evolution*, **37**, 163–179.

Bernatchez, L and Wilson, CC (1998). Comparative phylogeography of Nearctic and Palearctic fishes. *Molecular Ecology*, **7**, 431–452.

Beukeboom, LW and Vrijenhoek, RC (1998). Evolutionary genetics and ecology of sperm-dependent parthenogenesis. *Journal of Evolutionary Biology*, **11**, 755–782.

Bininda-Emonds, ORP, Gittleman, JL, and Purvis, A (1999). Building large trees by combining phylogenetic information: a complete phylogeny of the extant Carnivora (Mammalia). *Biological Reviews*, **74**, 143–175.

Blackburn, TM (1991). A comparative examination of lifespan and fecundity in parasitoid Hymenoptera. *Journal of Animal Ecology*, **60**, 151–164.

—— and Gaston, KJ (1994). The distribution of body sizes of the world's bird species. *Oikos*, **70**, 127–130.

—— and —— eds. (2003). *Macroecology: concepts and consequences*. Blackwell, Oxford.

Bleiweiss, R (1998). Origin of hummingbird faunas. *Biological Journal of the Linnean Society*, **65**, 77–97.

Böhm, M and Mayhew, PJ (2005). Historical biogeography and the latitudinal gradient of species richness in the Papionini (Primata: Cercopithecidae). *Biological Journal of the Linnean Society*, **85**, 235–246.

Bouton, N, Witte, F, van Alphen, JJM, Schenk, A, and Seehausen, O (1999). Local adaptations in populations of rock-dwelling haplochromines (Pisces: Cichlidae) from southern Lake Victoria. *Proceedings of the Royal Society of London, Series B*, **266**, 355–360.

Bronstein, J (2001). Mutualisms. In CW Fox, DA Roff, and DJ Fairburn, eds. *Evolutionary ecology: Concepts and case studies*, pp. 315–330. Oxford University Press, New York.

Brown, JH, (1995). *Macroecology*. Chicago University Press, Chicago, IL.

—— Marquet, PA, and Taper, ML (1993). Evolution of body size: consequences of an energetic definition of fitness. *American Naturalist*, **142**, 573–584.

—— and West, GB, eds. (2000). *Scaling in Biology*. Oxford University Press, Oxford.

Brown, VK and Southwood, TRE (1983). Trophic diversity, niche breadth, and the generation times of exopterygote insects in a secondary succession. *Oecologia*, **56**, 220–225.

Buchner, P (1965). *Endosymbiosis of animals with plant microorganisms*. Wiley Interscience, New York.

Bull, JJ (1983). *Evolution of sex determining mechanisms*. Benjamin/Cummings, Menlo Park.

—— (1994). Virulence. *Evolution*, **48**, 1423–1437.

Burgers, P and Chiappe, LM (1999). The wing of *Archaeopteryx* as a primary thrust generator. *Nature*, **399**, 60–62.

Burt, A and Trivers, R (1998). Genetic conflicts in genomic imprinting. *Proceedings of the Royal Society of London, Series B*, **265**, 2393–2397.

Capillon, C and Atlan, A (1999). Evolution of driving X chromosomes and resistance factors in experimental populations of *Drosophila simulans*. *Evolution*, **53**, 506–517.

Cardillo, M (1999). Latitude and rates of diversification in birds and butterflies. *Proceedings of the Royal Society of London, Series B*, **266**, 1221–1225.

Carlquist, S (1965). *Island life*. The Natural History Press, New York.

—— (1974). *Island biology*. Columbia University Press, New York.

Carroll, SB, Weatherbee, SD, and Langeland, JA (1995). Homeotic genes and the regulation and evolution of insect wing number. *Nature*, **375**, 58–61.

Caughley, G (1994). Directions in conservation biology. *Journal of Animal Ecology*, **63**, 215–244.

Charnov, EL (1976). Optimal foraging: the marginal value theorem. *Theoretical Population Biology*, **9**, 129–136.

—— (1982). *The theory of sex allocation*. Princeton University Press, Princeton.

—— (1991). Evolution of life history variation in female mammals. *Proceedings of the National Academy of Sciences, USA*, **88**, 1134–1137.

—— (1993). *Life history invariants: some explorations of symmetry in evolutionary ecology*. Oxford University Press, Oxford.

—— (2001). Evolution of mammal life histories. *Evolutionary Ecology Research*, **3**, 521–535.

—— (2004). The optimal balance between growth rate and survival in mammals. *Evolutionary Ecology Research*, **6**, 307–313.

—— and Bull, JJ (1977). When is sex environmentally determined? *Nature*, **266**, 828–830.

—— Los-den Hartogh, RL, Jones, WT and van den Assem, J (1981). Sex ratio evolution in a variable environment. *Nature*, **289**, 27–33.

—— Maynard Smith, J and Bull, JJ (1976). Why be an hermaphrodite? *Nature*, **263**, 125–126.

—— and Schaffer, WM (1973). Life history consequences of natural selection: Cole's result revisited. *American Naturalist*, **107**, 791–793.

Cleveland, LR (1947). The origin and evolution of meiosis. *Science*, **105**, 287–289.

Cody, ML and Overton, JMC (1996). Short-term evolution of reduced dispersal in island plant populations. *Journal of Ecology*, **84**, 53–61.

Cohen, D (1966). Optimizing reproduction in a randomly varying environment. *Journal of Theoretical Biology*, **12**, 119–129.

Connell, JH (1971). On the role of natural enemies in preventing competitive exclusion in some marine animals and in rain forest trees. In PJ Den Boer and G Gradwell, eds. *Dynamics of populations*, pp. 298–312. Pudoc, Wageningen.

Conover, DO and Munch, SB (2002). Sustaining fisheries yields over evolutionary time scales. *Science*, **297**, 94–96.

—— and Vanvorhees, DA (1990). Evolution of a balanced sex ratio by frequency-dependent selection in a fish. *Science*, **250**, 1556–1558.

Coope, GR (1973). Tibetan species of dung beetle from late-Pleistocene deposits in England. *Nature*, **245**, 335–336.

Cosmides, LM and Tooby, J (1981). Cytoplasmic inheritance and intragenomic conflict. *Journal of Theoretical Biology*, **89**, 83–129.

Crews, D (1994). Constraints to parthenogenesis. In RV Short and E Balaban, eds. *The differences between the sexes*, pp. 23–49. Cambridge University Press, Cambridge.

Cwynar, LC and MacDonald, GM (1987). Geographic variation in lodge-pole pine in relation to population history. *American Naturalist*, **129**, 463–469.

Davies, NB and Brooke, M de L (1989). An experimental study of coevolution between the cuckoo, *Cuculus canoris*, and its hosts, I and II. *Journal of Animal Ecology*, **58**, 207–236.

Darwin, C (1859). *On the origin of species by means of natural selection.* John Murray, London.

—— (1871). *The descent of man and selection in relation to sex.* John Murray, London.

de Haan, AA, Koelewijn, HP, Hundscheid, MPJ, and van Damme, JJM. (1997). The dynamics of gynodioecy in *Plantago lanceolata* L. II. Mode of action and frequencies of restorer genes. *Genetics*, **147**, 1317–1328.

de Kroon, H and Hutchings, MJ (1995). Morphological plasticity in clonal plants: the foraging concept revisited. *Journal of Ecology*, **83**, 143–152.

de Queiroz, A (1998). Interpreting sister-group tests of key innovation hypotheses. *Systematic Biology*, **47**, 710–718.

Denno, RF, Roderick, GK, Olmstead, KL, and Döbel, HG (1991). Density-related migration in planthoppers (Homoptera: Delphacidae): the role of habitat persistence. *American Naturalist*, **138**, 1513–1541.

Dial, KD (2003). Wing-assisted incline running and the evolution of flight. *Science*, **299**, 402–404.

Dial, KP and Marzluff, JM (1988). Are the smallest organisms the most diverse? *Ecology*, **69**, 1620–1624.

Dieckmann, U (1997). Can adaptive dynamics invade? *Trends in Ecology and Evolution*, **12**, 128–131.

—— and Doebeli, M. (1999). On the origin of species by sympatric speciation. *Nature*, **400**, 354–357.

Dudley, R (2000). The evolutionary physiology of animal flight. *Annual Review of Ecology and Systematics*, **62**, 135–155.

Dyer, LA (1995). Tasty generalists and nasty specialists? Antipredator mechanisms in tropical Lepidoptera larvae. *Ecology*, **76**, 1483–1496.

Dyson, EA and Hurst, GDD (2004). Persistence of an extreme sex-ratio bias in a natural population. *Proceedings of the National Academy of Sciences, USA*, **101**, 6520–6523.

Ebert, D and Herre, EA (1996). The evolution of parasitic diseases. *Parasitology Today*, **12**, 96–101.

Edwards, AVF (1998). Natural selection and the sex ratio: Fisher's sources. *American Naturalist*, **151**, 564–569.

Edwards, PJ, Kollmann, J, and Fleischmann, K (2002). Life history evolution in *Lodoicea maldivica* (Arecaceae). *Nordic Journal of Botany*, **22**, 227–237.

Ellner, S (1986). Germination dimorphisms and parent-offspring conflict in seed germination. *Journal of Theoretical Biology*, **123**, 173–185.

——(1987). Competition and dormancy—a reanalysis and review. *American Naturalist*, **130**, 798–803.

Enquist, BJ, Brown, JH, and West, GB (1998). Scaling of plant energetics and population density. *Nature*, **395**, 163–165.

——West, GB, Charnov, EL, and Brown, JH (1999). Allometric scaling of production and life history variation in vascular plants. *Nature*, **401**, 907–911.

Farrell, BD (1998). 'Inordinate fondness' explained: why are there so many beetles? *Science*, **281**, 555–559.

——Dussourd, DE, and Mitter, C (1991). Escalation of plant defenses: do latex and resin canals spur plant diversification? *American Naturalist*, **138**, 881–900.

Feder, JL (1998). The Apple Maggot Fly, *Rhagoletis pomonella*: flies in the face of conventional wisdom about speciation? In DJ Howard and SH Berlocher, eds. *Endless forms: species and speciation*, pp. 130–144. Oxford University Press, New York.

Fenchel, T and Finlay, BL (1995). *Ecology and evolution in anoxic worlds*. Oxford University Press, Oxford.

Fenner, F and Ratcliffe, FN (1965). *Myxomatosis*. Cambridge University Press, Cambridge.

Ferriere, R, Bronstein, JL, Rinaldi, S, Law, R, and Gauduchon, M (2002). Cheating and the evolutionary stability of mutualisms. *Proceedings of the Royal Society of London, Series B*, **269**, 773–788.

Fisher, DO and Owens, IPF (2004). The comparative method in conservation biology. *Trends in Ecology and Evolution*, **19**, 391–398.

Fisher, RA (1930). *The genetical theory of natural selection*. Oxford University Press, Oxford.

Foote, M (1996). Models of morphological diversification. In D Jablonski, DH Erwin, and JH Lipps, eds. *Evolutionary paleobiology*, pp. 62–86. University of Chicago Press, Chicago.

——(1997). The evolution of morphological diversity. *Annual Review of Ecology and Systematics*, **28**, 129–152.

Franco, M and Silvertown, J (1997). Life history variation in plants: an exploration of the fast-slow continuum hypothesis. In J Silvertown, M Franco, and JL Harper, eds. *Plant life histories: ecology, phylogeny and evolution*, pp. 210–227. Cambridge University Press, Cambridge.

Frank, SA (1996). Models of parasite virulence. *Quarterly Review of Biology*, **71**, 37–78.

—— (1997). Models of symbiosis. *American Naturalist*, **150** (Suppl.), S80–S99.

—— (2002). A touchstone in the study of adaptation. *Evolution*, **56**, 2561–2564.

Franklin, IR (1980). Evolutionary change in small populations. In ME Soulé and BA Wilcox, eds. *Conservation biology: an evolutionary-ecological perspective*, pp. 135–149. Sinauer, Sunderland.

Fretwell, SD and Lucas, HL (1970). On territorial behaviour and other factors influencing habitat distribution in birds. *Acta Biotheoretica*, **19**, 16–36.

Fryer, G and Iles, TD (1972). *The cichlid fishes of the Great Lakes of Africa: their biology and evolution*. Oliver and Boyd, Edinburgh.

Futuyma, DJ (2001). Ecological specialization and generalization. In CW Fox, DA Roff, and DJ Fairburn, eds. *Evolutionary ecology: concepts and case studies*, pp. 177–189. Oxford University Press, New York.

—— and Moreno, G (1988). The evolution of ecological specialization. *Annual Review of Ecology and Systematics*, **19**, 207–233.

Galis, F and Metz, JAJ (1998). Why are there so many cichlids? *Trends in Ecology and Evolution*, **13**, 1–2.

Ganeshaiah, KN and Uma Shaanker, R (1988). Seed abortion in wind-dispersed pods of *Dalbergia sissoo*: maternal regulation or sibling rivalry? *Oecologia*, **75**, 135–139.

Gardezi, T and da Silva, J (1999). Diversity in relation to body size in mammals: a comparative study. *American Naturalist*, **153**, 110–123.

Garner, JP, Taylor, GK, and Thomas, ALR (1999). On the origins of birds: the sequence of character acquisition in the evolution of avian flight. *Proceedings of the Royal Society of London, Series B*, **266**, 1259–1266.

Gaston, KJ and Blackburn, TM (1996). The tropics as a museum of biological diversity: an analysis of the New World avifauna. *Proceedings of the Royal Society of London, Series B*, **263**, 63–68.

—— and —— (1999). A critique for macroecology. *Oikos*, **84**, 353–368.

—— and —— (2000). *Pattern and process in macroecology*. Blackwell, Oxford.

Gavrilets, S (1999). Dynamics of clade diversification on the morphological hypercube. *Proceedings of the Royal Society of London, Series B*, **266**, 817–824.

—— (2003). Models of speciation: what have we learned in 40 years? *Evolution*, **57**, 2197–2215.

Gemmill, AW, Skorping, A, and Read, AF (1999). Optimal timing of first reproduction in parasitic nematodes. *Journal of Evolutionary Biology*, **12**, 1148–1156.

George, JC, Bada, J, Zeh, J *et al.* (1999). Age and growth estimates of bowhead whales (*Balaena mysticetus*) via aspartic acid racemization. *Canadian Journal of Zoology*, **77**, 571–580.

Gleeson, SK and Fry, JE (1997). Root proliferation and marginal patch value. *Oikos*, **79**, 387–393.

Godfray, HCJ (1994). *Parasitoids: behavioral and evolutionary ecology*. Princeton University Press, Princeton.

Goss-Custard, JD and Sutherland, WJ (1997). Individual behaviour, populations and conservation. In JR Krebs and NB Davies, eds. *Behavioural ecology: an evolutionary approach*, 4th edn, pp. 373–395. Blackwell, Oxford.

Grafen, A (1984). Natural selection, kin selection and group selection. In JR Krebs and NB Davies, eds. *Behavioural ecology, an evolutionary approach*, 2nd edn, pp. 62–84. Blackwell, Oxford.

Grant, BR and Grant, PR (1993). Evolution of Darwin's finches caused by a rare climatic event. *Proceedings of the Royal Society of London, Series B*, 251, 111–117.

Grant, PR, ed. (1998). *Evolution on islands*. Oxford University Press, Oxford.

——and Grant, BR (1996). Speciation and hybridization in island birds. *Philosophical Transactions of the Royal Society of London, Series B*, 351, 765–772.

Gyllenberg, M, Parvinen, K, and Dieckmann, U (2002). Evolutionary suicide and evolution of dispersal in structured metapopulations. *Journal of Mathematical Biology*, 45, 79–105.

Hahn, BH, Shaw, GM, de Cock, KM, and Sharp, PM (2000). Aids as a zoonosis: scientific and public health considerations. *Science*, 287, 607–614.

Haig, D and Westoby, M (1991). Genomic imprinting in endosperm: its effect on seed development in crosses between species, and between different ploidies of the same species, and its implications for the evolution of apomixis. *Proceedings of the Royal Society of London, Series B*, 333, 1–13.

Hamilton, WD (1963). The evolution of altruistic behaviour. *American Naturalist*, 97, 354–356.

——(1964a). The genetical evolution of social behaviour I. *Journal of Theoretical Biology*, 7, 1–16.

——(1964b). The genetical evolution of social behaviour II. *Journal of Theoretical Biology*, 7, 17–52.

——(1967). Extraordinary sex ratios. *Science*, 156, 477–488.

——(1980). Sex versus non-sex versus parasite. *Oikos*, 35, 282–290.

——(1996). *Narrow roads of gene land. Vol. 1 Evolution of social behaviour*. W.H. Freeman Spektrum, Oxford.

——and May, RM (1977). Dispersal in stable habitats. *Nature*, 269, 578–581.

Hanski, I and Ovaskainen, O (2000). The metapopulation capacity of a fragmented landscape. *Nature*, 404, 755–758.

Hardin, G (1968). The tragedy of the commons. *Science*, 162, 1243–1248.

Hardy, ICW, ed. (2002). *Sex ratios: concepts and research methods*. Cambridge University Press, Cambridge.

Harvey, PH, May, RM, and Nee, S (1994). Phylogenies without fossils. *Evolution*, 48, 523–529.

——Nee, S, Mooers, AO, and Partridge, L (1991). These hierarchical views of life: phylogenies and metapopulations. In RJ Berry, TJ Crawford, and GM Hewitt, eds. *Genes in Ecology*, pp. 123–137. Blackwell, Oxford.

——and Pagel, MD (1991). *The comparative method in evolutionary biology*. Oxford University Press, Oxford.

——and Purvis, A (1999). Understanding the ecological and evolutionary reasons for life history variation: mammals as a case study. In JM McGlade, ed. *Advanced Ecological Theory: Principles and Applications*, pp. 232–248. Blackwell, Oxford.

——and Zammuto, RM (1985). Patterns of mortality and age at first reproduction in natural populations of mammals. *Nature*, 315, 319–320.

Hawthorne, DJ and Via, S (2001). Genetic linkage of ecological specialization and reproductive isolation in pea aphids. *Nature*, **412**, 904–907.

Hedrick, PW (1986). Genetic polymorphism in heterogeneous environments: a decade later. *Annual Review of Ecology and Systematics*, **17**, 535–566.

—— (2001). Evolutionary conservation biology. In CW Fox, DA Roff, and DJ Fairburn, eds. *Evolutionary ecology: concepts and case studies*, pp. 371–383. Oxford University Press, New York.

—— and Kalinowski, ST (2000). Inbreeding depression in conservation biology. *Annual Review of Ecology and Systematics*, **31**, 139–161.

Heino, M (1998). Management of evolving fish stocks. *Canadian Journal of Fisheries and Aquatic Sciences*, **55**, 1971–1982

Herre, EA (1985). Sex ratio adjustment in fig wasps. *Science*, **228**, 896–898.

—— (1993). Population structure and the evolution of virulence in nematode parasites of fig wasps. *Science*, **259**, 1442–1445.

Herrero, M and Hormaza, JI (1996). Pistil strategies controlling pollen tube growth. *Sexual Plant Reproduction*, **9**, 343–347.

Hochberg, ME, Gomulkiewicz, R, Holt, RD, and Thompson JN (2000). Weak sinks could cradle mutualisms—strong sources should harbor pathogens. *Journal of Evolutionary Biology*, **13**, 213–222.

Hodges, SA and Arnold, ML (1995). Spurring plant diversification: are floral nectar spurs a key evolutionary innovation? *Proceedings of the Royal Society of London, Series B*, **262**, 343–348.

Holmes, EC (2001). On the origin and evolution of the human immunodeficiency virus (HIV). *Biological Reviews*, **76**, 239–254.

Howard, DJ and Berlocher, SH, eds. (1998). *Endless forms: species and speciation*. Oxford University Press, New York.

Hughes, CL, Hill, JK, and Dytham, C (2003). Evolutionary trade-offs between reproduction and dispersal in populations at expanding range boundaries. *Proceedings of the Royal Society of London, Series B*, **270**, S147–S150.

Hurst, LD and Hamilton, WD (1992). Cytoplasmic fusion and the nature of the sexes. *Proceedings of the Royal Society of London, Series B*, **247**, 189–194.

Hutchinson, GE (1959). Homage to Santa Rosalia, or why are there so many kinds of animals? *American Naturalist*, **93**, 245–249.

Hyatt, LA and Evans, AS (1998). Is decreased germination fraction associated with risk of sibling competition? *Oikos*, **83**, 29–35.

Imbert, E and Ronce, O (2001). Phenotypic plasticity for dispersal ability in the seed heteromorphic *Crepis sancta* (Asteraceae). *Oikos*, **93**, 126–134.

Jablonski, D (1986). Larval ecology and macroevolution of marine invertebrates. *Bulletin of Marine Science*, **39**, 565–587.

—— (1993). The tropics as a source of evolutionary novelty through geological time. *Nature*, **364**, 142–144.

Jaenike, J (1990). Host specialization in phytophagous insects. *Annual Review of Ecology and Systematics*, **21**, 243–273.

Janzen, DH (1970). Herbivores and the number of tree species in tropical forests. *American Naturalist*, **104**, 501–528.

Jeon, KW (1972). Development of cellular dependence on infective organisms: micrurgical studies in amoebas. *Science*, **176**, 1122–1123.

Johnson, ML and Gaines, MS (1990). Evolution of dispersal: theoretical models and empirical tests using birds and mammals. *Annual Review of Ecology and Systematics*, **21**, 449–480.

Johnson, MTJ and Agrawal, AA (2003). The ecological play of predator-prey dynamics in an evolutionary theatre. *Trends in Ecology and Evolution*, **18**, 549–551.

Jones, KE and Purvis, A (1997). An optimum body size for mammals? Comparative evidence from bats. *Functional Ecology*, **11**, 751–756.

Kanygin, AV (2001). The Ordovician phenomenon of explosive divergence of the Earth's organic realm: causes and effects on the biosphere evolution. *Geologiya I Geofizika*, **42**, 631–667.

Kawecki, TJ (1994). Accumulation of deleterious mutations and the evolutionary cost of being a generalist. *American Naturalist*, **144**, 833–838.

Kelly, CK (1990). Plant foraging: a marginal value model and coiling response in *Cuscuta subinclusa*. *Ecology*, **71**, 1916–1925.

Kiers, ET, Rousseau, RA, West, SA, and Denison, RF (2003). Host sanctions and the legume-rhizobia mutualism. *Nature*, **425**, 78–81.

Kindlmann, P, Dixon, AFG, and Dostálkova, I (1999). Does body size optimization result in skewed body size distribution on a logarithmic scale? *American Naturalist*, **153**, 445–447.

Kitchell, JA and Carr, TR (1985). Non equilibrium model of diversification: faunal turnover dynamics. In JW Valentine, ed. *Phanerozoic diversity patterns: profiles in macroevolution*, pp. 277–309. Princeton University Press and Pacific Division, AAAS, Princeton and San Fransisco.

Kocher, TD (2004). Adaptive evolution and explosive speciation: the cichlid fish model. *Nature Reviews Genetics*, **5**, 288–298.

Kondrashov, AS (1988) Deleterious mutations and the evolution of sexual reproduction. *Nature*, **336**, 435–440.

Kornfield, IL and Smith, PF (2000). African cichlid fishes: model systems for evolutionary biology. *Annual Review of Ecology and Systematics*, **31**, 163–196.

Kozlowski, J (2002). Theoretical and empirical status of Brown, Marquet and Taper's model of species-size distribution. *Functional Ecology*, **16**, 540–542.

—— and Gawelczyk, AT (2002). Why are species' body size distributions usually skewed to the right? *Functional Ecology*, **16**, 419–432.

—— and Konarzewski, M (2004). Is West, Brown and Enquist's model of allometric scaling mathematically correct and biologically relevant? *Functional Ecology*, **18**, 283–289.

—— Konarzewski, M, and Gawelczyk, AT (2003). Cell size as a link between noncoding DNA and metabolic rate scaling. *Proceedings of the National Academy of Sciences, USA*, **100**, 14080–14085.

—— and Weiner, J (1997). Interspecific allometries are by-products of body size optimization. *American Naturalist*, **149**, 352–379.

Krebs, JR and Davies, NB (1993). *An introduction to behavioural ecology* 3rd edn. Blackwell, Oxford.

Kukalová-Peck, J (1978). Origin and evolution of insect wings and their relationship to metamorphosis, as documented by the fossil record. *Journal of Morphology*, **156**, 53–126.

Kunin, WE and Gaston, KJ (1997). *The biology of rarity*. Chapman & Hall, London.

—— and Schmida, A (1997). Plant reproductive traits as a function of local, regional, and global abundance. *Conservation Biology*, **11**, 183–192.

Lande, R (1988). Genetics and demography in biological conservation. *Science*, **241**, 1455–1460.

—— (1993). Risk of population extinction from demographic and environmental stochasticity and random catastrophes. *American Naturalist*, **142**, 911–927.

—— (1995). Mutation and conservation. *Conservation Biology*, **9**, 782–791.

—— Seehausen, O, and van Alphen, JJM (2001). Mechanisms of rapid sympatric speciation by sex reversal and sexual selection in cichlid fish. *Genetica*, **112–113**, 435–443.

Law, R (2000). Fishing, selection and phenotypic evolution. *ICES Journal of Marine Science*, **57**, 659–668.

—— and Dieckmann, U (1998). Symbiosis through exploitation and the merger of lineages in evolution. *Proceedings of the Royal Society of London, Series B*, **265**, 1245–1253.

—— and Grey, DR (1989). Evolution of yields from populations with age-specific cropping. *Evolutionary Ecology*, **3**, 343–359.

Lawton, JH (1995). Population dynamic principles. In JH Lawton and RM May, eds. *Extinction rates*, pp. 147–163. Oxford University Press, Oxford.

Lazcano, A and Miller, SL (1999). On the origin of metabolic pathways. *Journal of Molecular Evolution*, **49**, 424–431.

Ledig, FT, Jacob-Cervantes, V, Hodgskiss, PD, and Eguiluz-Piedra T. (1997). Recent evolution and divergence among populations of a rare Mexican endemic, Chihuahua spruce, following Holocene climatic warming. *Evolution*, **51**, 1815–1827.

Lenton, TM, Schellnhuber, HJ, and Szathmáry, E (2004). Climbing the co-evolution ladder. *Nature*, **431**, 913.

Lessells, CM (1991). The evolution of life histories. In JR Krebs and NB Davies, eds. *Behavioural ecology: an evolutionary approach*, 3rd edn., pp. 32–68. Blackwell, Oxford.

Lessios, HA (1998). The first stage of speciation as seen in organisms separated by the Isthmus of Panama. In DJ Howard and SH Berlocher, eds. *Endless forms: species and speciation*, pp. 186–201. Oxford University Press, New York.

Levin, DA (2000). *The origin, expansion and demise of plant species*. Oxford University Press, Oxford.

Levins, R (1968). *Evolution in changing environments*. Princeton University Press, Princeton.

Liem, KF (1973). Evolutionary strategies and morphological innovations: cichlid pharyngeal jaws. *Systematic Zoology*, **22**, 425–441.

Lively, CM (2001). Parasite–host interactions. In CW Fox, DA Roff, and DJ Fairburn, eds. *Evolutionary ecology: concepts and case studies*, pp. 290–302. Oxford University Press, New York.

Logan, GA, Hayes, JM, Hieshima, GB, and Summons, RE (1995). Terminal Proterozoic reorganization of biogeochemical cycles. *Nature*, **376**, 53–56.

MacArthur, RH and Pianka, ER (1966). On optimal use of a patchy environment. *American Naturalist*, **100**, 603–609.

Majerus, MEN (2003). *Sex wars: genes, bacteria, and biased sex ratios*. Princeton University Press, Princeton.

Marden, JH and Kramer, MG (1995). Locomotor performance of insects with rudimentary wings. *Nature*, **377**, 332–334.

Margulis, L and Bermudes, D (1985). Symbiosis as a mechanism of evolution: status of cell symbiosis theory. *Symbiosis*, **1**, 101–124.

Maurer, BA (1998a). The evolution of body size in birds. I. Evidence for non-random diversification. *Evolutionary Ecology*, **12**, 925–934.

——(1998b). The evolution of body size in birds. II. The role of reproductive power. *Evolutionary Ecology*, **12**, 935–944.

——Brown, JH, and Rusler, RD (1992). The micro and macro in body size evolution. *Evolution*, **46**, 939–953.

Mayhew, PJ (1997). Adaptive patterns of host-plant selection by phytophagous insects. *Oikos*, **79**, 417–428.

——(2002). Shifts in hexapod diversification and what Haldane could have said. *Proceedings of the Royal Society of London, Series B*, **269**, 969–974.

——and Blackburn, TM (1999). Does development mode organize life history evolution in the parasitoid Hymenoptera? *Journal of Animal Ecology*, **68**, 906–916.

——and Glaizot, O (2001). Integrating theory of clutch size and body size evolution for parasitoids. *Oikos*, **92**, 372–376.

——and Hardy, ICW (1998). Nonsiblicidal behavior and the evolution of clutch size in bethylid wasps. *American Naturalist*, **151**, 409–424.

Maynard Smith, J (1966). Sympatric speciation. *American Naturalist*, **100**, 637–650.

——(1979). The effect of normalizing and disruptive selection on genes for recombination. *Genetical Research*, **33**, 121–128.

——(1984). The ecology of sex. In JR Krebs and NB Davies, eds. *Behavioural ecology: an evolutionary approach*, 2nd edn., pp. 201–210. Blackwell, Oxford.

——and Szathmáry, E (1995). *The major transitions in evolution*. Freeman, Oxford.

——and——(1999). *The origins of life: from the birth of life to the origin of language*. Oxford University Press, Oxford.

Mayr, E (1940). Speciation phenomena in birds. *American Naturalist*, **74**, 249–278.

——(1963). *Animal species and evolution*. Harvard University Press, Cambridge.

McKinnley, ML (1990). Trends in body size evolution. In KJ McNamara, ed. *Evolutionary trends*, pp. 75–118. University of Arizona Press, Tucson.

McLaughlin, JF, Hellmann, JJ, Boggs, CL, and Ehrlich, PR (2002). Climate change hastens population extinctions. *Proceedings of the National Academy of Sciences, USA*, **99**, 6070–6074.

Meyer, A (1993). Phylogenetic relationships and evolutionary processes in African cichlids. *Trends in Ecology and Evolution*, **8**, 279–284.

Miller, B and Mullette, KJ (1985). Rehabilitation of an endangered Australian bird: Lord Howe Woodhen *Tricholimnas sylvestris*. *Biological Conservation*, **34**, 55–95.

Mitter, C, Farrell, B, and Wiegmann, B (1988). The phylogenetic study of adaptive zones: has phytophagy promoted insect diversification? *American Naturalist*, **132**, 107–128.

Mock, DW and Parker, GA (1997). *The evolution of sibling rivalry*. Oxford University Press, Oxford.

Morse, DR, Lawton, JH, Dodson, MM, and Williamson, MH (1985). Fractal dimensions of vegetation and the distribution of arthropod body lengths. *Nature*, **314**, 731–733.

Myers, N, Mittermeier, RA, Mittermeier, CG, da Fonseca, GAB, and Kent J (2000). Biodiversity hotspots for conservation priorities. *Nature*, **403**, 853–858.

Nee, S, Barraclough, TG, and Harvey, PH (1996). Temporal changes in biodiversity: detecting patterns and identifying causes. In KJ Gaston, ed. *Biodiversity: A biology of numbers and difference*, pp. 230–252. Oxford University Press, Oxford.

Nowak, MA, Anderson, RM, McLean, AR, Wolfs, T, Goudsmit, J, and May, RM (1991). Antigenic diversity thresholds and the development of AIDS. *Science*, **254**, 963–969.

Orme, CDL, Isaac, NJB, and Purvis, A. (2002). Are most species small? Not within species-level phylogenies. *Proceedings of the Royal Society of London, Series B*, **269**, 1279–1287.

Orr, HA (1998). Testing natural selection versus genetic drift in phenotypic evolution using Quantitative Trait Locus data. *Genetics*, **149**, 2099–2104.

Owens, IPF and Bennett, PM (2000). Ecological basis of extinction risk in birds: habitat loss versus human persecution and introduced predators. *Proceedings of the National Academy of Sciences, USA*, **97**, 12144–12148.

—————— and Harvey, PH (1999). Species richness among birds: body size, life history, sexual selection or ecology? *Proceedings of the Royal Society of London, Series B*, **266**, 933–939.

Padian, K and Chiappe, LM (1998). The origin and early evolution of birds. *Biological Reviews*, **73**, 1–42.

Pake, CE and Venable, DL (1996). Seed banks in desert annuals: implications for persistence and coexistence in variable environments. *Ecology*, **77**, 1427–1435.

Parker, GA and Maynard Smith, J (1990). Optimality theory in evolutionary biology. *Nature*, **348**, 27–33.

Pellmyr, O and Huth, CJ (1994). Evolutionary stability of mutualism between yuccas and yucca moths. *Nature*, **372**, 257–260.

Perrin, N and Mazalov, V (2000). Local competition, inbreeding and the evolution of sex-biased dispersal. *American Naturalist*, **155**, 116–127.

Policansky, D (1987). Evolution, sex and sex allocation. *Bioscience*, **37**, 466–468.

Prinzing, A (2003). Are generalists pressed for time? An interspecific test of the time-limited disperser model. *Ecology*, **84**, 1744–1755.

Promislow, DEL and Harvey, PH (1990). Living fast and dying young: a comparative analysis of life history variation among mammals. *Journal of Zoology*, **220**, 417–437.

Purvis, A (1996). Using interspecies phylogenies to test macroevolutionary hypotheses. In PH Harvey, AJ Leigh Brown, J Maynard Smith, and S Nee, eds. *New uses for new phylogenies*, pp. 153–168. Oxford University Press, Oxford.

Purvis, A and Harvey, PH (1995). Mammal life history evolution: a comparative test of Charnov's model. *Journal of Zoology*, **237**, 259–283.

—— Jones, KE, and Mace, GM (2000). Extinction. *Bioessays*, **22**, 1123–1133.

—— Nee, S, and Harvey, PH (1995). Macroevolutionary influences from primate phylogeny. *Proceedings of the Royal Society of London, Series B*, **260**, 329–333.

Queller, DC (1983). Sexual selection in a hermaphroditic plant. *Nature*, **305**, 706–708.

Ralls, K, Brugger, K, and Ballou, J (1979). Inbreeding and juvenile mortality in small populations of ungulates. *Science*, **206**, 1101–1103.

Ranius, T and Hedin, J (2001). The dispersal rate of a beetle, *Osmoderma eremita*, living in tree hollows. *Oecologia*, **126**, 363–370.

Rees, M (1993). Trade-offs among dispersal strategies in the British Flora. *Nature*, **366**, 150–152.

—— (1997). Evolutionary ecology of seed dormancy and seed size. In J Silvertown, M Franco, and JL Harper, eds. *Plant life histories: ecology, phylogeny and evolution*, pp. 121–142. Cambridge University Press, Cambridge.

Ricklefs, RE and O'Rourke, K (1975). Aspect diversity in moths: a temperate-tropical comparison. *Evolution*, **29**, 313–324.

—— and Renner, SS (2000). Evolutionary flexibility and flowering plant diversity: a comment on Dodd, Silvertown, and Chase. *Evolution*, **54**, 1061–1065.

Rieseberg, LH, Raymond, O, Rosenthal, DM *et al.* (2003). Major ecological transitions in annual sunflowers facilitated by hybridization. *Science*, **301**, 1211–1216.

—— VanFossen, C, and Desrochers, AM (1995). Hybrid speciation accompanied by genomic reorganization in wild sunflowers. *Nature*, **375**, 313–316.

—— Whitton, J, and Gardner, K (1999). Hybrid zones and the genetic architecture of a barrier to gene flow between two wild sunflower species. *Genetics*, **152**, 713–727.

Rigaud, T (1997). Inherited microorganisms and sex determination of arthropod hosts. In SL O'Neill, AA Hoffman, and JH Werren, eds. *Influential passengers: inherited microorganisms and arthropod reproduction*, pp. 81–102. Oxford University Press, Oxford.

Roff, DA (1990). The evolution of flightlessness in insects. *Ecological Monographs*, **60**, 389–421.

—— (1992). *The evolution of life histories: theory and analysis*. Chapman & Hall, New York.

—— (1994). The evolution of flightlessness: is history important? *Evolutionary Ecology*, **8**, 629–657.

Roughgarden, J (1995). *Anolis lizards of the Caribbean: ecology, evolution and plate tectonics*. Oxford University Press, Oxford.

—— and Pacala, SW (1989). Taxon cycle among *Anolis* lizard populations: review of the evidence. In D Otte and J Endler, eds. *Speciation and its consequences*, pp. 403–432. Sinauer, Sunderland.

Roy, K and Foote, M (1997). Morphological approaches to measuring biodiversity. *Trends in Ecology and Evolution*, **12**, 277–281.

——Jablonski, D, and Martien, KK (2000). Invariant size-frequency distributions along a latitudinal gradient in marine bivalves. *Proceedings of the National Academy of Sciences, USA*, **97**, 13150–13155.

Rundle, HD, Mooers, AØ, and Whitlock, MC (1998). Single founder-flush events and the evolution of reproductive isolation. *Evolution*, **52**, 1850–1855.

Saccheri, I, Kuussaari, M, Kankare, M, Vikman, P, Fortelius, W, and Hanski, I (1998). Inbreeding and extinction in a butterfly metapopulation. *Nature*, **392**, 491–494.

Saumitou-Laprade, P, Cuguen, J, and Vernet, P (1994). Cytoplasmic male sterility in plants: molecular evidence and the nucleocytoplasmic conflict. *Trends in Ecology and Evolution*, **9**, 431–435.

Schluter, D (1998). Ecological causes of speciation. In DJ Howard and SH Berlocher, eds. *Endless forms: species and speciation*, pp. 114–129. Oxford University Press, New York.

——(2000). *The ecology of adaptive radiation*. Oxford University Press, Oxford.

——(2001). Ecological character displacement. In CW Fox, DA Roff, and DJ Fairburn, eds. *Evolutionary ecology: concepts and case studies*, pp. 265–276. Oxford University Press, New York.

Schilthuizen, M (2001). *Frogs, flies and dandelions: the making of species*. Oxford University Press, Oxford.

Searcy, KB and Macnair, MR (1993). Developmental selection in response to environmental conditions of the maternal parent in *Mimulus guttanus*. *Evolution*, **47**, 13–24.

Seehausen, O (2000). Explosive speciation rates and unusual species richness in haplochromine cichlid fishes: effects of sexual selection. *Advances in Ecological Research*, **31**, 237–274.

——Mayhew, PJ, and van Alphen, JJM (1999a). Evolution of colour patterns in East African cichlid fish. *Journal of Evolutionary Biology*, **12**, 514–534.

——and van Alphen, JJM (1998). The effect of male coloration on female mate choice in closely related Lake Victoria cichlids (*Haplochromis nyererei* complex). *Behavioural Ecology and Sociobiology*, **42**, 1–8.

——and ——(1999). Can sympatric speciation by disruptive sexual selection explain rapid evolution of cichlid diversity in Lake Victoria? *Ecology Letters*, **2**, 262–271.

——— and Lande, R (1999b). Color polymorphism and sex ratio distortion in a cichlid fish as an incipient stage in sympatric speciation by sexual selection. *Ecology Letters*, **2**, 367–378.

——— and Witte, F (1997). Cichlid fish diversity threatened by eutrophication that curbs sexual selection. *Science*, **277**, 1808–1811.

Seger, J and Stubblefield, JW (2002). Models of sex ratio evolution. In Hardy, ICW, ed. *Sex ratios: concepts and research methods*, pp. 2–25. Cambridge University Press, Cambridge.

Sepkoski, JJ, Jr (1999). Rates of speciation in the fossil record. In AE Magurran and RM May, eds. *Evolution of biological diversity*, pp. 260–282. Oxford University Press, Oxford.

Sereno, PC (1999). The evolution of dinosaurs. *Science*, **284**, 2137–2147.

Shorrocks, B, Marsters, J, Ward, I, and Evennett, PJ (1991). The fractal dimension of lichens and the distribution of arthropod body lengths. *Functional Ecology*, 5, 457–460.

Signor, PW (1990). The geological history of diversity. *Annual Review of Ecology and Systematics*, 21, 509–539.

Silvertown, J and Gordon, DM (1989). A framework for plant behaviour. *Annual Review of Ecology and Systematics*, 20, 349–366.

Simpson, GG (1953). *The major features of evolution*. Columbia University Press, New York.

Singer, MC, Ng, D, and Thomas, CD (1988). Heritability of oviposition preference and its relationship to offspring performance within a single insect population. *Evolution*, 42, 977–985.

——Thomas, CD, and Parmesan, C (1993). Rapid human-induced evolution of insect-host associations. *Nature*, 366, 681–683.

Skogsmyr, I and Lankinen, Å (2000). Female assessment of good genes in stylar tissue. *Evolutionary Ecology Research*, 2, 965–979.

——and —— (2002). Sexual selection: an evolutionary force in plants? *Biological Reviews*, 77, 537–562.

Slatkin, M (1980). Ecological character displacement. *Ecology*, 61, 163–177.

Solignac, M and Monnerot, M (1986). Race formation, speciation, and introgression within *Drosophila simulans*, *D. mauritiana*, and *D. sechellia* inferred from mitochondrial DNA analysis. *Evolution*, 40, 531–539.

Soulé, ME (1987). *Viable populations for conservation*. Cambridge University Press, Cambridge.

Southwood, TRE (1962). Migration of terrestrial arthropods in relation to habitat. *Biological Reviews*, 37, 171–214.

——(1978). The components of diversity. In Mound LANW, ed. *Diversity of insect faunas*, pp. 19–40. Blackwell, Oxford.

Southwood, TRE (2003). *The story of life*. Oxford University Press, Oxford.

Speakman, JR (2001). The evolution of flight and echolocation in bats: another leap in the dark. *Mammal Review*, 31, 111–130.

Stanley, SM (1979). *Macroevolution: pattern and process*. Freeman, San Fransisco.

Stearns, SC, ed. (1987). *The evolution of sex and its consequences*. Birkauser, Basel.

——(1992). *The evolution of life histories*. Oxford University Press, Oxford.

Stehli, FG, Douglas, RG, and Kafescioglu, IA (1972). Models for the evolution of planktonic foraminifera. In Schopf, TJM, ed. *Models in paleobiology*, pp. 116–128. Freeman, Cooper and Co, San Fransisco.

Steiner, KE and Whitehead, VB (1990). Pollinator adapation to oil-secreting flowers—*Rediviva* and *Diascia*. *Evolution*, 44, 1701–1707.

Stephenson, AG and Winsor, JA (1986). *Lotus corniculatus* regulates offspring quality through selective fruit abortion. *Evolution* 40: 453–458.

Stillman, RA, Goss-Custard, JD, West, AD *et al.* (2000). Predicting mortality in novel environments: tests and sensitivity of a behaviour-based model. *Journal of Applied Ecology*, 37, 564–588.

Stouthamer, R, Luck, RF, and Hamilton, WD (1990). Antibiotics cause parthenogenetic *Trichogramma* (Hymenoptera/Trichogrammatidae) to revert to sex. *Proceedings of the National Academy of Sciences, USA*, **87**, 2424–2427.

Sutherland, S (1987). Why hermaphroditic plants produce many more flowers than fruits: experimental tests with *Agave mckelveyana*. *Evolution*, **41**, 750–759.

Sutherland, WJ (1996). Predicting the consequences of habitat loss for migrating populations. *Proceedings of the Royal Society of London, Series B*, **263**, 1325–1327.

—— and Norris, K (2002). Behavioural models of population growth rates: implications for conservation and prediction. *Philosophical Transactions of the Royal Society of London, Series B*, **357**, 1273–1284.

—— and Stillman, RA (1988). The foraging tactics of plants. *Oikos*, **52**, 239–244.

Symonds, MRE (1999). Insectivore life histories: further evidence against an optimum body size for mammals. *Functional Ecology*, **13**, 508–513.

Szathmáry, E (1993). Do deleterious mutations act synergistically? Metabolic control theory provides a partial answer. *Genetics*, **133**, 127–132.

—— and Maynard Smith, J (1995). The major evolutionary transitions. *Nature*, **374**, 227–232.

Taper, ML and Case, TJ (1985). Quantitative genetic model for the coevolution of character displacement. *Ecology*, **66**, 355–371.

Thomas, ALR and Norberg, RA (1996). Skimming the surface—the origin of flight in insects? *Trends in Ecology and Evolution*, **11**, 187–188.

Thompson, JN (1987). Symbiont-induced speciation. *Biological Journal of the Linnean Society*, **32**, 385–393.

—— (1989). Concepts of coevolution. *Trends in Ecology and Evolution*, **4**, 179–183.

—— (1994). *The coevolutionary process*. University of Chicago Press, Chicago.

—— (1998). Rapid evolution as an ecological process. *Trends in Ecology and Evolution*, **13**, 329–332.

—— (1999). The raw material for coevolution. *Oikos*, **84**, 5–16.

—— (2001). The geographic dynamics of coevolution. In CW Fox, DA Roff, and DJ Fairburn, DJ, eds. *Evolutionary ecology: concepts and case studies*, pp. 331–343. Oxford University Press, New York.

—— and Burdon, JJ (1992). Gene-for-gene coevolution between plants and parasites. *Nature*, **360**, 121–125.

Towe, KM (1990). Aerobic respiration in the Archaen? *Nature*, **348**, 54–56.

Travis, JJM and Dytham, C (1999). Habitat persistence, habitat availability, and the evolution of dispersal. *Proceedings of the Royal Society of London, Series B*, **266**, 723–728.

—— and —— (2002). Dispersal evolution during invasions. *Evolutionary Ecology Research*, **4**, 1119–1129.

Trivers, RL and Willard, DE (1973). Natural selection of parental ability to vary the sex ratio of offspring. *Science*, **179**, 90–92.

Turkington, R (1989). The growth, distribution and neighbour relationships of *Trifolium repens* in a permanent pasture. V. The coevolution of competitors. *Journal of Ecology*, **77**, 717–733.

Turner, GF (1999). Explosive speciation of African cichlid fishes. In AE Magurran and RM May, eds. *Evolution of biological diversity*, pp. 113–129. Oxford University Press, Oxford.

Vamosi, JC, Otto, SP, and Barrett, SCH (2003). Phylogenetic analysis of the ecological correlates of dioecy in angiosperms. *Journal of Evolutionary Biology*, 16, 1006–1018.

van Doorn, GS, Noest, AJ, and Hogeweg, P (1998). Sympatric speciation and extinction driven by environment dependent sexual selection. *Proceedings of the Royal Society of London, Series B*, 265, 1915–1919.

van Doorn, GS and Weissing, FJ (2001). Ecological versus sexual selection models of sympatric speciation: a synthesis. *Selection*, 2, 17–40.

Van Valen, L (1973). A new evolutionary law. *Evolutionary Theory*, 1, 1–30.

Venable, DL and Brown, JS (1988). The selective interactions of dispersal, dormancy and seed size as adaptations for reducing risk in variable environments. *American Naturalist*, 131, 360–384.

Verhulst, P-F (1845). Recherches mathematiques sur la loi d'accroisse- ment de la population. *Nouveaux Memoires de l'Academie Royale des Sciences, des Lettres et des Beaux-Arts de Belgique*, 18, 1–32.

Vermeij, GJ (1995). Economics, volcanoes, and phanerozoic revolutions. *Paleobiology*, 21, 125–152.

Via, S (2001). Sympatric speciation in animals: the ugly duckling grows up. *Trends in Ecology and Evolution*, 16, 381–390.

Vrijenhoek, RC (1994). Unisexual fish. *Annual Review of Ecology and Systematics*, 25, 71–96.

Wächtershauser, G (1988). Before enzymes and templates: theory of surface metabolism. *Micobiological Reviews*, 52, 452–484.

——(1990). Evolution of the first metabolic cycles. *Proceedings of the National Academy of Sciences, USA*, 87, 200–204.

Warren, MS, Hill, JK, Thomas, JA *et al.* (2001). Rapid responses of British butterflies to opposing forces of climate and habitat change. *Nature*, 414, 65–69.

Waxman, D and Gavrilets, S (2005). Twenty questions on adaptive dynamics. *Journal of Evolutionary Biology*, 18, 1139–1154.

Werren, JH (1983). Sex ratio evolution under local mate comptetition in a parasitic wasp. *Evolution*, 37, 116–124.

—— and Beukeboom, LW (1998). Sex determination, sex ratios and genetic conflict. *Annual Review of Ecology and Systematics*, 29, 233–261.

West, GB, Brown, JH, and Enquist, BJ (1999). The fourth dimension of life; fractal geometry and allometric scaling of organisms. *Science*, 284, 1677–1679.

—— —— and —— (2001). A general model for ontogenetic growth. *Nature*, 413, 628–631.

——Woodruff, WH, and Brown, JH (2002). Allometric scaling of metabolic rate from molecules and mitochondria to cells and mammals. *Proceedings of the National Academy of Sciences, USA*, 99, 2473–2478.

West, SA and Herre, EA (2002). Using sex ratios: why bother? In ICW Hardy, ed. *Sex ratios: concepts and research methods*, pp. 399–413. Cambridge University Press, Cambridge.

Whitham, TG (1980). The theory of habitat selection examined and extended using *Pemphigus* aphids. *American Naturalist*, **115**, 449–466.

Whitlock, MC (1996). The red queen beats jack-of-all-trades: the limitations on the evolution of phenotypic plasticity and niche breadth. *American Naturalist*, **148** (Suppl), S65–S77.

Wijesinghe, DK and Hutchings, MJ (1996). Consequences of patchy distribution of light for the growth of the clonal herb *Glechoma hederacea*. *Oikos*, **77**, 137–145.

Wilkinson, DM and Sherratt, TN (2001). Horizontally acquired mutualisms, an unsolved problem in ecology? *Oikos*, **92**, 377–384.

Willig, MR, Kaufman, DM, and Stevens, RD (2003). Latitudinal gradients of biodiversity: pattern, process, scale, and synthesis. *Annual Review of Ecology and Systematics*, **34**, 272–309.

Willis, KJ and McElwain, JC (2002). *The evolution of plants*. Oxford University Press, Oxford.

Wynne-Edwards, VC (1962). *Animal dispersion in relation to social behaviour*. Hafner, New York.

Xiong, J, Fischer, WM, Inoue, K, Nakahara, M, and Bauer, CE. (2000). Molecular evidence for the early evolution of photosynthesis. *Science*, **289**, 1724–1730.

Xu, X, Zhou, ZH, Wang, XL, Kuang, XW, Zhang, FC, and Du, XK (2003). Four-winged dinosaurs from China. *Nature*, **421**, 335–340.

Yamamura, N, Higashi, M, Behera, N, and Wakano, JY (2004). Evolution of mutualism through spatial effects. *Journal of Theoretical Biology*, **226**, 421–428.

Young, TP (1990). Evolution of semelparity in Mount Kenya Lobelias. *Evolutionary Ecology*, **4**, 157–172.

Yule, GU (1924). A mathematical theory of evolution based on the conclusions of Dr, JC Willis FRS. *Philosophical Transactions of the Royal Society of London, Series A*, **213**, 21–87.

# Glossary

Abiotic—not living.

Alleles—variant forms of a gene at a particular **locus**.

Allozymes—identifiable forms of an **enzyme** coded for by **alleles** at a single **locus**.

Ammonoids—an extinct group of squid-like marine molluscs with coiled shells.

Anthers—the male parts of a flower, releasing pollen.

Apterygotes—primitively wing-less insects, including silverfish and bristletails.

Archaebacterium—a member of a diverse group of bacteria differing biochemically from other 'Eubacteria', and today living in extreme environments, such as anaerobic mud, animal guts, salt lakes, and hot springs.

Archosaur—a member of the Archosauria, a group of terrestrial vertebrates including crocodiles, birds, pterosaurs, and dinosaurs.

Autosomes—**chromosomes** that are not sex chromosomes.

Autotrophic—capable of utilizing inorganic carbon as the main source of carbon and of obtaining energy for life processes from the oxidation of inorganic elements (chemotrophic) or from radiant energy (phototrophic).

Benthic—living on the bottom of a water body.

Biosphere—the components of planet Earth, including water, soil, and atmosphere, in which organisms may be found.

Biotic—living.

Bivalves—a class of mollusc, including clams, oysters, cockles and their kin, surrounded by a pair of hinged shells.

Bottlenecks—temporary reductions in population size that reduce the genetic diversity of a population.

Carrying capacity—the density above which a population cannot be expected to increase.

Cell cycle—the processes between the birth of a cell by division, and its growth and subsequent division into daughter cells.

Cell membrane—the layer surrounding and containing the fluid part of a cell.

Cell wall—a non-living structure outside the **cell membrane** of plants, fungi, and bacteria that provides support and protection for cells.

Centriole—the **organelle** that forms the spindle fibres that separate the **chromosomes** during cell division.

Cephalopods—a class of marine molluscs, including squid, octopi, cuttlefish, nautili, as well as the extinct ammonoids and belemnoids.

Chemo-autotrophs—**Autotrophic** bacteria that can synthesize organic compounds from inorganic raw materials in the absence of sunlight. The energy is derived from the oxidation of inorganic materials, such as hydrogen sulphide, ammonia, and iron-bearing compounds.

Chlorophyll—the green pigment in plants that absorbs sunlight and uses its energy to synthesize carbohydrates from $CO_2$ and water, the process of **photosynthesis**.

Chloroplasts—**eukaryotic organelles**, originally derived from **cyanobacteria**, and that carry out **photosynthesis**.

Chromatids—The daughter strands of a duplicated **chromosome** joined together at a structure called the centromere.

Chromosomes—a structure consisting of a very long piece of **DNA** carrying many genes.

Clade—a group of organisms that share a single common ancestor, representing a single branch of the evolutionary tree.

Cladistic revolution—a series of rapid scientific advances for estimating the evolutionary tree of life.

**Co-evolutionary**—involving co-evolution, the reciprocal evolution of interacting species.

**Cosexual**—an individual that expresses both gender functions, male and female, within its lifetime.

**Cursorial**—running on the ground.

**Crinoids**—a group of marine animals, related to sea urchins and starfish, containing sea lilies and feather stars.

**Cyanobacteria**—a group of photosynthetic bacteria, some of which became the **chloroplasts** of plants.

**Cytoplasmic**—in or of the cytoplasm, the fluid interior of a cell.

**Cytoskeleton**—a three-dimensional structure within the fluid interior of **eukaryotic** cells that aids movement and stability.

**Detritivores**—organisms that feed on decomposing organic material.

**Dioecious**—when only one sex is expressed in a single individual.

**Diploid**—cells containing **chromosomes** in **homologous** pairs, one derived from each parent.

**Directional selection**—selection favouring individuals with extreme traits (e.g. larger than average) within a population, such that the average trait shifts over time.

**DNA**—the acronym for deoxyribonucleic acid, the molecule that contains the genetic information within cells.

**Echinoids**—starfish, sea urchins, and their kin.

**Effective population size**—the size of an ideal population which acts the same, with respect to the degree of **genetic drift** or inbreeding, as the real population in question. Effective population sizes are usually smaller than real populations because of variation in reproductive success among individuals or deviation from a 50 : 50 sex ratio.

**Enzymes**—**proteins** that speed up biochemical reactions within cells, acting as **metabolic catalysts**.

**Eukaryotes**—organisms distinct from bacteria, which contain a **nucleus** containing non-circular **chromosomes**, a **cytoskeleton**, and other **organelles**, such as **mitochondria** and **chloroplasts**.

**Eutherians**—placental mammals, distinct from monotremes and marsupials which both lack a placenta.

**Extant**—still living, the opposite of extinct.

**Foraminifera**—single-celled aquatic **protists** with shells that are a prominent component of marine ecosystems and of the fossil record.

**Functional groups**—groups of species that perform a similar ecological role.

**Gametes**—the **haploid** cells, sperm, and egg that fuse to form a **diploid** cell in a sexual life cycle.

**Gastropods**—snails and their relatives.

**Genetic drift**—an evolutionary process, more important in small populations, in which the genetic composition of a population changes through random events.

**Genetic load**—the extent to which the average individual in a population is less fit than the fittest individual. This equals the relative chance that an average individual will die before reproducing because of the deleterious **alleles** that it possesses.

**Genome**—the total genetic content of a cell.

**Geochemical cycles**—the movement and transformation of materials around the Earth.

**Graptolites**—extinct, stick-like, colonial marine organisms, found in the early Palaeozoic. Possibly related to group of worms known as hemichordates.

**Haploid**—cells containing just one member of each **homologous chromosome** pair.

**Heterotrophic**—organisms obtaining their carbon and energy from organic sources.

**Homonoids**—the group containing gibbons, great apes, and humans.

**Homologous**—alike due to shared ancestry. Homologous **chromosomes** in a cell derived from different parents but contain equivalent sets of genes.

**Horizontally transmitted**—transmitted between unrelated individuals within a single generation as well as across generations, as opposed to vertical transmission (parent to offspring).

**Inbreeding coefficient**—The probability that a **zygote** obtains copies of the same ancestral **allele** from both its parents because the parents are related to each other.

**Leguminous plants**—plants from the family Leguminosae, containing peas, beans, and their relatives, which form a symbiotic association with **rhizobia** bacteria in their root nodules.

**Limnetic**—the main water column of a lake.

**Linkage**—the tendency for certain genes to be inherited together because they are located on the same **chromosome**.

**Locus**—the position on a **chromosome** where a particular gene is located.

**Macroscopic**—visible with the naked eye, as opposed to microscopic.

**Meiotic**—to do with meiosis, the form of cell division in which a **diploid** cell gives rise to four **haploid** cells.

**Mendelian inheritance**—the form of inheritance, discovered by Gregor Mendel, which occurs in most sexual organisms, whereby each individual inherits two copies of a gene, one from each parent.

**Metabolic catalyst**—a molecule that facilitates biochemical reactions within cells.

**Metabolic cycle**—the flow of material through a series of biochemical reactions within an organism.

**Metapopulation**—a population consisting of many ephemeral subpopulations, which are linked by dispersal between them. The metapopulation persists by replacing extinct subpopulations through recolonization.

**Mitochondria**—**eukaryotic organelles**, originally derived from a group of bacteria, which perform aerobic respiration within cells.

**Mitotic**—involving mitosis, a type of cell division in which a (normally diploid) cell gives rise to two (normally diploid) daughter cells.

**Molecular revolution**—a series of rapid scientific advances in understanding the biochemical basis of biology, and particularly in reading the genetic code.

**Monoplacophora**—a primitive class of mollusc, today living in deep-ocean trenches.

**Mycorrhizal fungi**—fungi that form a symbiosis by living within the roots of plants.

**Neutral evolution**—evolution that has no effect on individual fitness, mainly consisting of changes to non-coding parts of the **genome** or synonymous changes to coding elements.

**Nuclear**—in or of the nucleus, the membrane-surrounded structure in **eukaryotic** cells that contains the **chromosomes**.

**Organelles**—the distinctive structural elements within a cell.

**Ostracods**—a class of aquatic crustacean, also called seed shrimps.

**Parasitoids**—insects, mainly wasps and flies, which develop as juveniles by feeding on the body of another host organism, usually another insect.

**Phanerozoic**—the time period beginning about 535 Ma, consisting of the Palaeozoic, Mesozoic, and Cenozoic, in which fossils of animals are abundant.

**Phenological**—to do with phenology, the timing of biological events.

**Phenotype**—the physical parts of an organism, as opposed to the genotype, which is the heritable blueprint for creating the phenotype, encoded in a cell's **DNA**.

**Photosynthesize**—carry out photosynthesis, the process by which some organisms synthesize organic molecules from inorganic carbon using sunlight.

**Phytoplankton**—the **photosynthetic** component of plankton.

**Pistils**—the central part of a flower, containing the female reproductive parts.

**Pleiotropy**—when one gene influences two or more **phenotypic** traits.

**Polymorphism**—when individuals in a single species can be of two or more distinct **phenotypes**.

**Population genetics**—the field that deals with how genes influence the characteristics of a population, particularly the processes that determine the frequency and distribution of **alleles**.

**Power function**—when a number is multiplied by itself a specified number of times. The number that specifies how many times is known as the exponent.

**Prokaryotes**—bacteria, organisms without a **nucleus, cytoskeleton, mitochondria**, or **chloroplasts**, and with a single circular **chromosome**. As opposed to **eukaryotes**.

**Protein**—a type of organic molecule, many of which function as **enzymes**, created from an **RNA** template in a process known as translation.

**Protist**—**eukaryotes**, often single-celled, which are not plants, animals, or fungi.

**Pterygotes**—the winged insects.

**Quantitative genetics**—the field concerned with measurable, continuous, or 'quantitative' traits and their evolution, which attempts, in particular, to predict the response to selection of a population.

**Raptorial**—predatory. When applied to birds, implies a member of the Falconiformes (vultures, falcons, hawks, eagles).

**Rhizobia**—a group of nitrogen-fixing bacteria that form a symbiotic association in the roots of **leguminous plants**.

**RNA**—the acronym for ribonucleic acid, a single-stranded molecule formed from a **DNA** template via a process called transcription, and involved in **protein** synthesis.

**Stabilizing selection**—selection against the extremes of a population, which results in the population average staying the same.

**Substitution**—changes to the **DNA** sequence of an organism (mutations) in which one nucleotide base (letter) is replaced by another different base (letter).

**Sympatric speciation**—speciation in which the two incipient daughter species have identical or substantially overlapping geographic ranges.

**T-lymphocytes**—vertebrate white blood cells involved in the immune response and which target particular foreign or cancer cells.

**Trilobites**—A group of extinct marine arthropods, characterized by a three-lobed body, which were a dominant component of the marine fauna in the Early Palaeozoic.

**Unisexual**—a species which has lost one of its sexes and become parthenogenetic through the surviving sex.

**Zooplankton**—the animal component of plankton.

**Zygote**—a fertilized egg.

# Index